Essential Soil Science

Essential Soil Science

Simon Miller

MURPHY & MOORE
www.murphy-moorepublishing.com

Essential Soil Science
Simon Miller
ISBN: 978-1-63987-203-9 (Hardback)

MURPHY & MOORE
Published by Murphy & Moore Publishing,
1 Rockefeller Plaza,
New York City, NY 10020, USA

Cataloging-in-Publication Data

Essential soil science / Simon Miller.
p. cm.
Includes bibliographical references and index.
ISBN 978-1-63987-203-9
1. Soil science. 2. Soils. 3. Agriculture. I. Miller, Simon.
S591 .E87 2022
631.4--dc23

For more information regarding Murphy & Moore Publishing and its products, please visit the publisher's website www.murphy-moorepublishing.com

Table of Contents

Preface

Soil is a substance which consists of minerals, organic matter, living organisms, gas, and water. Soil science studies the soil as a natural resource that is found on Earth. It is divided into two branches, namely, pedology and edaphology. While pedology focuses on formation, chemistry, morphology and classification of soil, edaphology is concerned with the interaction of soil on living things. Concepts from this discipline are used for solving problems related to escalating population growth, water crisis and increased land degradation have made soil science an important subject. This book unfolds the innovative aspects of soil science which will be crucial for the holistic understanding of the subject matter. While understanding the long-term perspectives of the topics, it makes an effort in highlighting their impact as a modern tool for the growth of the discipline. This book will serve as a valuable source of reference for those interested in this field.

A foreword of all chapters of the book is provided below:

Chapter 1- The complex mixture of water, air, minerals, organic matter and several living organisms which makes a thin layer covering the surface of the Earth is known as soil. It supports plant life which is vital for survival. This is an introductory chapter which will introduce briefly all the significant aspects of soil such as formation of soil and weathering, soil forming factors, soil composition, soil structure, soil grade, etc. **Chapter 2**- The branch of science which is concerned with the study of soil and its various aspects including its classification, formation and its physical, chemical and biological properties is known as Soil science. It has various branches like soil physics, soil chemistry and soil biology. The topics elaborated in this chapter will help in gaining a better perspective about the branches of soil science. **Chapter 3**- Soil is rich in various nutrients which are essential for the growth of plants such as nitrogen, potassium and phosphorus. This chapter includes soil fertility, essential elements, macronutrients, micronutrients, sodium in soil, soil ion exchange soil pH, soil acidity, etc. All the diverse principles of soil nutrition have been carefully analyzed in this chapter. **Chapter 4**- Nutrient cycling in soil refers to putting back the nutrients of soil into biological circulation by liberating carbon and nutrients from complex materials through decomposition. This chapter discusses in detail ion absorption by plant roots and carbon, nitrogen, phosphorus and potassium cycle in soil. The chapter closely examines the key concepts of soil nutrient cycle to provide an extensive understanding of the subject. **Chapter 5**- Pedology is a scientific discipline which deals with the formation and evolution of soil. In order to completely understand pedology, it is necessary to understand the processes related to it. The following chapter elucidates the varied processes and mechanisms associated with this area of study such as pedogenesis, soil morphology, active layer and soil map. **Chapter 6**- The branch of soil science which includes the ecological study of soil and its influence on living things, especially plants is known as edaphology. This chapter covers topics like agricultural soil science, environmental soil science, bioeffector, plant nutrition and phosphate solubilizing bacteria. The aim of this chapter is to explore the field of edaphology. **Chapter 7**-The systematic study of description, classification and mapping of soil is known as soil survey. It applies the theories of soil formation, physical geography, land use patterns and analysis of vegetation. This chapter closely examines the key concepts of soil survey including catena, liming and pedotransfer function, etc. to provide an extensive understanding of the subject. **Chapter 8**- The presence of contaminants in soil beyond a certain concentration which cause loss of functions of soil is known as soil contamination. The transportation of upper layer of soil from one place on Earth's surface to another place due to the action of water, wind, animals or human activities is known as soil erosion. This chapter has been carefully written to provide an easy understanding of the varied facets of soil loss and soil contamination. **Chapter 9**- There are various tools and techniques used in soil science. One of them is soil moisture sensor. It is a device which is used to measure the water content present in the soil. This chapter encompasses the tools and techniques related to soil science such as soil nailing, soil compaction, soil bioengineering and pedometrics. The topics elaborated in this chapter will help in gaining a better perspective about these techniques.

I would like to make a special mention of my publisher who considered me worthy of this opportunity and also supported me throughout the process. I would also like to thank the editing team at the back-end who extended their help whenever required.

Simon Miller

1

Soil: Formation, Structure and Chemical Composition

The complex mixture of water, air, minerals, organic matter and several living organisms which makes a thin layer covering the surface of the Earth is known as soil. It supports plant life which is vital for survival. This is an introductory chapter which will introduce briefly all the significant aspects of soil such as formation of soil and weathering, soil forming factors, soil composition, soil structure, soil grade, etc.

The word soil is derived from word 'solum' which means earthly material in which plants grow. The study of soils as naturally occurring phenomena is called pedology (from the Greek word pedon, meaning soil or earth). Soil possesses complexity and heterogeneity. It documentation is as early as 5000 BC. Vedas and Upanishads regarded soil as 'Mother of all' that supports and nourishes all life on Earth. By definition, Soil is a dynamic natural body developed as a result of pedogenic processes during and after weathering of rocks, consisting of mineral and organic constituents, possessing definite chemical, physical, mineralogical and biological properties, having a variable depth over the surface of the earth and providing a medium for plant growth for land plants.

Types of Soil

Soil can be defined in many ways. Soil is a naturally occurring, loose/un-cemented/weakly cemented/relatively unconsolidated mineral particles, organic or inorganic in character, lying over the bed rock which is formed by weathering of rocks. Soil is formed by different particles such as gravel, rock, sand, silt, clay, loam and humus.

Sandy Soil

Sandy Soil is a soil with a high percentage of sand, or large soil particles. It mainly consists of rock particles such as limestone, shale, granite and quartz. Sandy Soil has a low percentage of organic material. Water travels through sandy soil very easily, so nutrients leach out quickly.

Characteristics

Sandy soils, those with 85% to 90% sand-sized particles, can be difficult to manage. They don't hold water well and they are not very fertile. Excessively sandy soils are characteristically gritty to the touch. Their large pores make it just as easy for water to run out as in. Nutrients also easily slip in and out of sandy soils.

On the plus side, these soils are not prone to erosion by water, since the water goes right through them. But they can be easily eroded by wind, depending on their location and prevailing winds. Sandy soils are easy to cultivate. They don't get too wet and they don't get sticky like heavy, clay soils. They can be worked right after a rain without difficulty, and they can be worked when they are dry.

Uses

Sandy soil is useful for mixing in for increasing drainage in clay beds or by itself when needing high drainage substrates. When gardening, some plants love it but those that require a lot of nutrients, it must be supplemented with some sort of organic material like compost, humus, or coconut coir. Interestingly enough, compost doesn't necessarily add very much in the way of nutrients on its own. It helps the soil retain nutrients from other fertilizers like guano, manure, etc. Humus and coir have no real nutrient content at all.

How to Manage Sandy Soil

Because sandy soil doesn't hold on to either water or mineral nutrients very well, it's important to apply them frequently. Check the soil often and water as soon as it is just barely moist. It can go from moist to very dry quickly, so be alert. Farmers call this type of soil "droughty". Sandy soil absorbs water quickly, so runoff is seldom a problem. Because it drains so quickly, letting ample air into the soil, overwatering is seldom a problem either.

But here's a caveat: a few sandy soils have just the right mixture of silt-sized particles to have very low porosity. The silt plugs up the pores between sand particles so water doesn't enter it quickly, and it has drainage problems. This soil is extremely dense but, because it doesn't contain clay, has no structure to speak of. Handle it as you would any other sandy soil, by adding organic matter. In a short time, the organic matter will open the soil and improve drainage. In the meantime, apply water slowly to avoid puddling.

Soluble fertilizers wash out of sandy soil quickly, so it's important to use slow-release fertilizers. Look for fertilizers with large percentages (more than 50% of the nitrogen) of water-insoluble nitrogen (WIN). Nitrogen in water-insoluble forms must be acted on by soil bacteria before it is available to plants. This takes a while, so the nutrients are released over a period of weeks instead of all at once.

Some plants have evolved on sandy soils, and are able to cope with dry periods without being harmed. Select plants that are suited to the soil you have as well as the climate you live in.

How to Improve Sandy Soil

Adding organic matter—or clay and organic matter—to sandy soils is the only way to increase their ability to retain water and nutrients. The organic matter is decomposed by soil bacteria into humus, a soft, sticky material that holds both water and nutrients like a sponge.

Humus is more effective in the presence of clay. If your soil has very little clay content, add some at the same time as the organic matter. Mix in about one inch of clay soil one time only. You don't need to add it often, as you do with organic matter.

Add organic matter as often as you can. Work in 1 to 4 inches every time you till a vegetable garden or annual bed, or prepare soil for planting. Keep a mulch of organic matter under trees, shrubs, and perennial flowers. Earthworms and insects will work it into the soil for you.

Compost, peat moss, well-rotted manure, or other easily available organic matters are good choices for working into sandy soils or for mulches. Mulches that decompose slowly, like bark chunks, will help somewhat, but their action is slow. Rapidly-decomposing mulches like compost and manure are more effective at improving the soil.

Farmers preparing sandy soil for planting acacia tree seedlings, Senegal.

Soil management practices which lead to an increase in the fine fraction are helpful in improving soil properties and crop productivity.

- Fertilization of these soils is considered essential. Inorganic fertilization is the main practice.
- Application of organic manures can supply nutrients in slowly available forms and improve soil physico–chemical properties.
- Surface application of organic manures to sandy soils does not last so the manure should be dug deeper into the soil or a carpet-like layer spread of not less than one centimeter thick, which will improve water storage, biological activity, nutrient status and increase yields.
- Mulch can be added to improved water storage by reducing evaporation. Crop residues, on the surface of the soil reduce evaporation losses, decrease the range between maximum and minimum soil temperature, and reduce wind erosion.
- For tillage to be really effective, it has to be done at the earliest possible time after irrigation or rainfall when the evaporation rate is still high.
- Minimum tillage, maintenance of a cover crop, strip cropping, crop rotations, control of grazing and establishment of shelter belts and windbreaks are some of the protective measures to counter the high susceptibility of sandy soils to erosion.
- Besides the conventional dry vegetation method, use of artificial surface sealants such as petroleum, synthetic rubber, chemicals and water soluble plastics have also been adopted for dune and drift sand stabilization.
- Afforestation with selected trees and shrubs is a complementary measure that should follow stabilization of dunes.

- Overgrazing on coarse textured soils must be avoided. The introduction of rotational grazing helps to combat this hazard. It might be better not to permit grazing but to use fodder cut on feeding lots.

Silt Soil

Silt Soil is a light and moisture retentive soil type with a high fertility rating. As silt soils compromise of medium sized particles they are well drained and hold moisture well. As the particles are fine, they can be easily compacted and are prone to washing away with rain. By adding organic matter, the silt particles can be bound into more stable clumps.

Clay Soil

Clay soil is soil that is comprised of very fine mineral particles and not much organic material. The resulting soil is quite sticky since there is not much space between the mineral particles, and it does not drain well at all.

Clay soil is prevalent many parts of the United States, and it can be very problematic if you are trying to grow a flower or vegetable garden. While some trees and shrubs grow well in clay, most annuals, perennials, and vegetables don't have roots strong enough to force their way through. dense clay. And if spring flower bulbs are your dream, forget it—most bulbs tend to rot over the winter in clay soils. Clay soils can be improved, however. With some background information and a well-designed strategy, you'll be able to grow flowers and vegetables to your heart's content.

If you have noticed that water tends to puddle on the ground rather than soak in, it is likely your ground consists of clay. Soil that consists of over 50 percent clay particles is referred to as "heavy clay." To determine whether you have clay soil or not, you can do a simple soil test. But chances are you probably already know if you have clay soil. If your soil sticks to shoes and a garden tool like glue, forms big clods that aren't easy to separate, and crusts over and cracks in dry weather, you have clay.

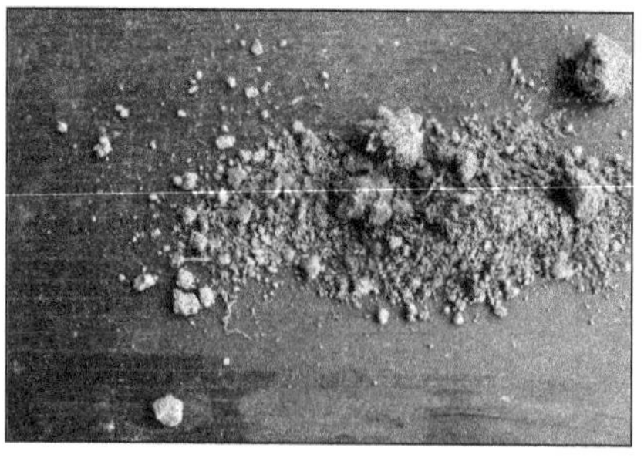

Advantages of Clay Soil

Even clay soil has some good qualities. Clay, because of its density, retains moisture well. It also tends to be more nutrient-rich than other soil types. The reason for this is that the particles that make up clay soil are negatively charged, which means they attract and hold positively charged particles, such as calcium, potassium, and magnesium.

Disadvantages of Clay Soil

Clay also has the following negative qualities:

- Slow draining.
- Slow to warm in the spring.
- Compacts easily, making it difficult for plant roots to grow.
- Tendency to heave in winter.
- Tendency to be alkaline in pH.

Improving Clay Soil

Improving your clay soil will take a bit of work, but the good news is that the work you do will instantly improve the structure of your soil and make it easier to work with. Most of the work is done up front, although some annual chores are necessary to prolong the soil improvement.

It is best to improve an entire planting area all at once, rather than to attempt improving the soil in individual planting holes as you need them. If you dig a planting hole in clay soil, then drop in a plant and nicely amend only the soil you're using to backfill, your plant will be happy for a little while. But you have essentially done nothing more than create an in-ground flower pot. Eventually, the plant will start sending out roots that will be stopped in their tracks when they reach the clay walls of the planting hole. You'll end up with a root-bound plant that it won't grow as large or as healthy as it should.

Start by defining the growing area for your garden bed. If you are improving an existing bed, you can dig out any plants you want to keep and set them aside in pots until your soil improvement is completed.

To improve your soil, you'll need to add 6 to 8 inches of organic matter to the entire bed. You can add any organic matter you can get your hands on. Grass clippings (as long as they haven't been treated with chemicals), shredded leaves, rotted manure, and compost are all perfect choices. Spread your organic matter on top of the soil. Here's where the manual labor comes in. The organic matter needs to be mixed into the top 6 to 12 inches of soil. Digging it in and mixing it with a shovel is a great way to do this, as it moves a lot of earth without pulverizing the soil particles the way tilling can. However, if digging is just too hard on your back, using a tiller is a fine method.

When you're finished, your garden bed will be several inches higher than it was originally, but this is not a problem. Your garden bed will settle some over the course of a season as the organic material breaks down. The soil structure will continue to improve as microorganisms in the soil work to break down all of the organic matter you've added.

The bed can be planted immediately. Plan to add more organic matter in the form of compost once or twice a year. This will continue the process of improving the soil's structure and offset any settling that happens.

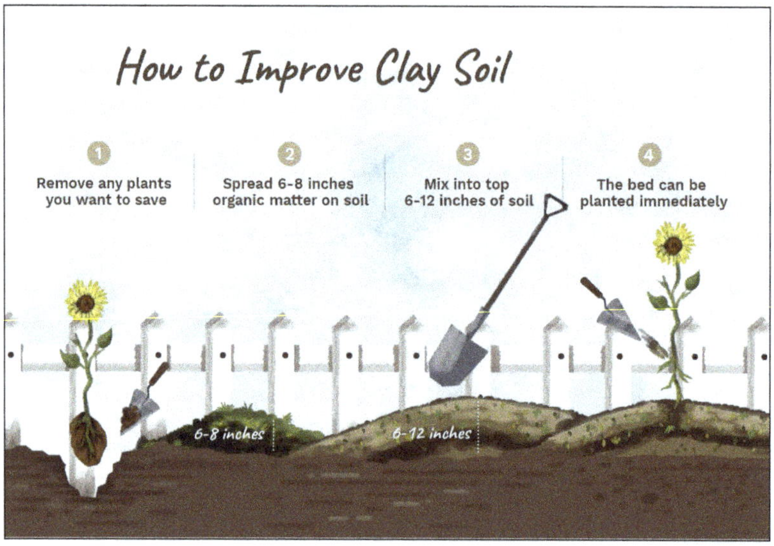

What not to Do

Surprisingly often, people imagine that the proper way to improve a dense, clay soil is to add the opposite kind of mineral material—sand. After all, loamy soils, viewed as ideal garden soil, are a mixture of sand and clay. Unfortunately, when sand is added directly to clay, the result is something that more accurately resembles concrete. The reason loamy soils are great for plants is that they have a large ratio of organic material in them as a foundation for the sand and clay. Without lots of organic material, clay plus sand equals an awful garden. Clay soils are best improved with the addition of compost and other organic materials only.

Sandy Loam Soil

Sandy loam soils are dominated by sand particles, but contain enough clay and sediment to provide some structure and fertility. There are four different types of sandy loam soil that are classified based on the size of the sand particles in the soil.

Classification

Sandy loam soils are broken down into four categories, including coarse sandy loam, fine sandy loam, sandy loam and very fine sandy loam. The size of the sand particles is measured in millimeters and their concentration in the soil is used to determine which category a soil falls under. Sandy loam soils are made of approximately 60 percent sand, 10 percent clay and 30 percent silt particles.

Characteristics

Sandy loam soils have visible particles of sand mixed into the soil. When sandy loams soils are compressed, they hold their shape but break apart easily. Sandy loam soils have a high concentration of sand that gives them a gritty feel. In gardens and lawns, sandy loam soils are capable of quickly draining excess water but cannot hold significant amounts of water or nutrients for your plants. Plants grown in this type of soil will require more frequent irrigation and fertilization than soils with a higher concentration of clay and sediment. Sandy loam soils are often deficient in specific micronutrients and may require additional fertilization to support healthy plant growth.

Identification

You can quickly identify sandy loam soil based on its physical characteristics. Pick up a handful of dry soil and slowly dribble water onto it. Work the water into the soil with your hand until it has a smooth consistency similar to putty. Hold the soil in your hand as though you are holding a pipe straight up and down and squeeze it. Sandy loam soils have a very gritty texture. If your soil is a sandy loam, it will form a cohesive ribbon of soil as it squeezes out between your thumb and finger that will fall apart before it reaches one inch in length.

Considerations

Plants that are grown in a sandy loam soil need frequent irrigation and fertilization to maintain healthy growth. The best way to improve a sandy loam soil for gardening is to mix organic matter into the soil. Incorporating a 2- to 4-inch layer of compost or peat moss over the area can significantly improve the ability of your sandy loam soil to hold nutrients and water.

Formation of Soil

Soil is formed as a result of long term process of complex interactions disintegration and decomposition of rocks due to weathering leading to the production of mineral matrix in close association with interstitial organic matter. Two types of rocks that are recognized are (1) Soft rock that is formed from decomposition of rocks and soil development proceed simultaneously and (2) in hard rock decomposition of rocks and rock minerals occurs prior to development of soil.

Soils evolve under the action of biological, climatic, geologic, and topographic influences. The evolution of soils and their properties is called soil formation, and pedologists have identified five fundamental soil formation processes that influence soil properties. These five "state factors" are parent material, topography, climate, organisms, and time.

Parent Material

Parent material is the initial state of the solid matter making up a soil. It can consist of consolidated rocks, and it can also include unconsolidated deposits such as river alluvium, lake or marine sediments, glacial tills, loess (silt-sized, wind-deposited particles), volcanic ash, and organic matter (such as accumulations in swamps or bogs). Parent materials influence soil formation through their mineralogical composition, their texture, and their stratification (occurrence in layers). Dark-coloured ferromagnesian (iron- and magnesium-containing) rocks, for example, can produce soils with a high content of iron compounds and of clay minerals in the kaolin or smectite groups, whereas light-coloured siliceous (silica-containing) rocks tend to produce soils that are low in iron compounds and that contain clay minerals in the illite or vermiculite groups. The coarse texture of granitic rocks leads to a coarse, loamy soil texture and promotes the development of E horizons (the leached lower regions of the topmost soil layer). The fine texture of basaltic rocks, on the other hand, yields soils with a loam or clay-loam texture and hinders the development of E horizons. Because water percolates to greater depths and drains more easily through soils with coarse texture, clearly defined E horizons tend to develop more fully on coarse parent material.

Leptosol soil profile from Switzerland, showing a typically shallow surface horizon with little evidence of soil formation.

In theory, parent material is either freshly exposed solid matter (for example, volcanic ash immediately after ejection) or deep-lying geologic material that is isolated from atmospheric water and organisms. In practice, parent materials can be deposited continually by wind, water, or volcanoes and can be altered from their initial, isolated state, thereby making identification difficult. If a single parent material can be established for an entire soil profile, the soil is termed monogenetic; otherwise, it is polygenetic. An example of polygenetic soils is soils that form on sedimentary rocks or unconsolidated water- or wind-deposited materials. These so-called stratified parent materials can yield soils with intermixed geologic layering and soil horizons—as occurs in south-eastern England, where soils forming atop chalk bedrock layers are themselves overlain by soil layers formed on both loess and clay materials that have been modified by dissolution of the chalk below:

Andosol soil profile from Italy, showing a dark-coloured surface horizon derived from volcanic parent material.

Adjacent soils frequently exhibit different profile characteristics because of differing parent materials. These differing soil areas are called lithosequences, and they fall into two general types. Continuous lithosequences have parent materials whose properties vary gradually along a transect, the prototypical example being soils formed on loess deposits at increasing distances downwind from their alluvial source. Areas of such deposits in the central United States or China show systematic decreases in particle size and rate of deposition with increasing distance from the source. As a result, they also show increases in clay content and in the extent of profile development from weathering of the loess particles.

By contrast, discontinuous lithosequences arise from abrupt changes in parent material. A simple example might be one soil formed on schist (a silicate-containing metamorphic rock rich in mica) juxtaposed with a soil formed on serpentine (a ferromagnesian metamorphic rock rich in olivine). More subtle discontinuous lithosequences, such as those on glacial tills, show systematic variation of mineralogical composition or of texture in unconsolidated parent materials.

Topography

Topography, when considered as a soil-forming factor, includes the following: the geologic structural characteristics of elevation above mean sea level, aspect (the compass orientation of a landform), and slope configuration (i.e., either convex or concave), and relative position on a slope (that is, from the toe to the summit). Topography influences the way the hydrologic cycle affects earth material, principally with respect to runoff processes and evapotranspiration. Precipitation may run off the land surface, causing soil erosion, or it may percolate into soil profiles and become part of subsurface runoff, which eventually makes its way into the stream system. Erosive runoff is most likely on a convex slope just below the summit, whereas lateral subsurface runoff tends to cause an accumulation of soluble or suspended matter near the toe slope. The conversion of precipitation into evapotranspiration is favoured by lower elevation and an equatorially facing aspect.

Adjacent soils that show differing profile characteristics reflecting the influence of local topography are called toposequences. As a general rule, soil profiles on the convex upper slopes in a toposequence are shallower and have less distinct subsurface horizons than soils at the summit or on lower, concave-upward slopes. Organic matter content tends to increase from the summit down to the toeslope, as do clay content and the concentrations of soluble compounds.

Often the dominant effect of topography is on subsurface runoff (or drainage). In humid temperate regions, well-drained soil profiles near a summit can have thick E horizons (the leached layers) overlying well-developed clay-rich Bt horizons, while poorly drained profiles near a toeslope can have thick A horizons overlying extensive Bg horizons (lower layers whose pale colour signals stagnation under water-saturated conditions). In humid tropical regions with dry seasons, these profile characteristics give way to less distinct horizons, with accumulation of silica, manganese, and iron near the toeslope, whereas in semiarid regions soils near the toeslope have accumulations of the soluble salts sodium chloride or calcium sulfate.

These general conclusions are tempered by the fact that topography is susceptible to great changes over time. Soil erosion by water or wind removes A horizons and exposes B horizons to weathering. Major portions of entire soil profiles can move downslope suddenly by the combined action of water and gravity. Catastrophic natural events, such as volcanic eruptions, earthquakes, and devastating storms, can have obvious consequences for the instability of geomorphologic patterns.

Climate

The term climate in pedology refers to the characteristics of weather as they evolve over time scales longer than those necessary for soil properties to develop. These characteristics include precipitation, temperature, and storm patterns—both their averages and their variation.

Cryosol soil profile from Canada showing patterned deformations caused by cycles of freeze and thaw.

Climate influences soil formation primarily through effects of water and solar energy. Water is the solvent in which chemical reactions take place in the soil, and it is essential to the life cycles of soil organisms. Water is also the principal medium for the erosive or percolative transport of solid particles. The rates at which these water-mediated processes take place are controlled by the amount of energy available from the sun.

On a global scale, the integrated effects of climate can readily be seen along a transect from pole to Equator. As one proceeds from the pole to cool tundra or forested regions, polar desert soils give way to intensively leached soils such as the Podzols (Spodosols) that exhibit an eye-catching ash-coloured E horizon indicative of humid boreal climates. Farther into temperate zones, organic matter accumulates in soils as climates become warmer, and eventually lime (calcium carbonate) also begins to accumulate closer to the top of the soil profile as evapotranspiration increases. Arid subtropical climate then follows, with desert soils that are low in organic matter and enriched in soluble salts. As the climate again becomes humid close to the Equator, high temperature combines with high precipitation to create red and yellow tropical soils, whose colours reveal the prevalence of residual iron oxide minerals that are resistant to leaching losses because of their low solubility.

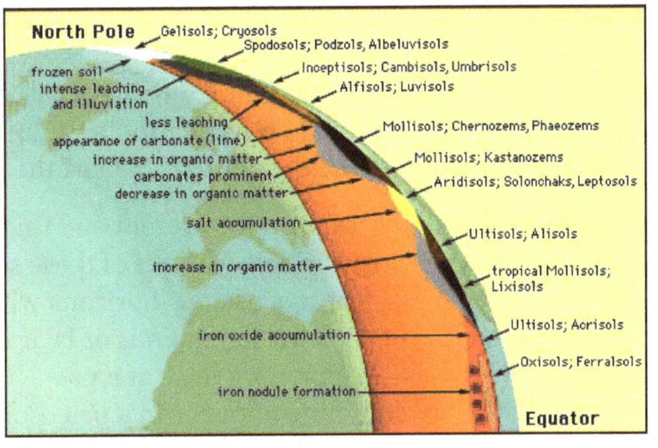

Climate And Soil: The effects of climate on soil worldwide. A schematic cross section of Earth illustrates the variation in soil types arising from differences in climate from the North Pole to the Equator.

On a continental scale, a transect taken across the central United States from east to west shows the effects of increasing evapotranspiration. First, soils that exhibit E horizons appear, followed by soils high in organic matter. These give way to soils with accumulations of lime and ultimately to desert soils with soluble salt efflorescence (powdery crust) near the surface.

On a regional scale, variations in climate also can influence soil properties significantly, resulting in a contiguous array of soils called a climosequence. One typical climosequence occurs along a 1,000-km (600-mile) north-south transect through the foothills of the Cascade and Sierra Nevada mountains in California. There soils that have formed on landscapes of similar topography vary continuously in their profile characteristics with variations in annual precipitation. Soils formed at the dry southern end of the transect are shallow and rocky, whereas those at the humid northern end show well-developed B horizons and reddish colour. Clay mineralogy in the upper 20 cm (8 inches) of these soils also responds to the increase in precipitation, shifting from the smectite group to the mixed vermiculite or illite group/kaolin group and finally to the kaolin group alone. These changes result primarily from increasing loss of silica and soluble metals as soil leaching extends deeper with increasing rainfall. In addition, soil acidity and organic matter content increase, while readily soluble forms of calcium (important to plant growth and soil aggregation) decrease, with increasing precipitation.

In principle, soil profile characteristics that are closely linked to climate can in turn be interpreted as climatic indicators. For example, a soil profile with two well-defined zones of lime accumulation, one shallow and one deep, may signal the existence of a past climate whose greater precipitation drove the lime layer deeper than the present climate is able to do.

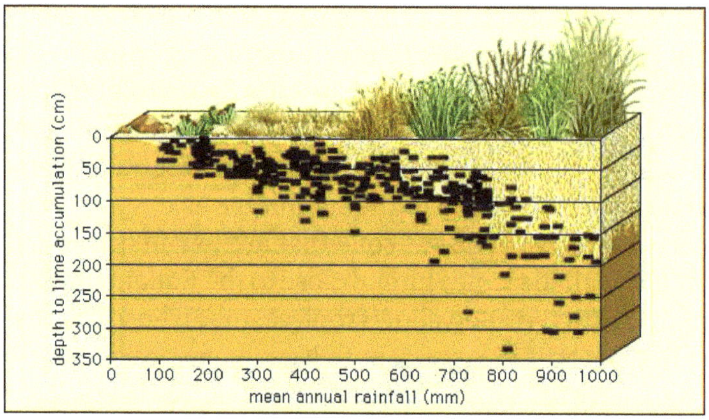

Lime Deposits.

Depth to lime accumulation in relation to annual rainfall. Lime ($CaCO_3$) deposits that can prevent the penetration of plant roots are found deeper in the soil profile in climates with higher mean annual rainfall than in climates where there is little water to transport the lime through the soil.

Soils that formed in past environments different from the present and that are preserved (at least partially) at greater depth are known as paleosols. Some features of these soils can serve as climatic indicators, the most reliable being robust features such as horizons with hardened accumulations of relatively insoluble iron, manganese, or calcium minerals or layers with accumulations of strongly aggregated clay-size particles. The precipitation level of a past climate might be inferred from an observation of the depth of lime-containing horizons in a paleosol. These potential applications of climatic relationships must be evaluated carefully in order to distinguish the effects of previously weathered parent material from those of purely climatic influence.

Organisms

The development of soils can be significantly affected by vegetation, animal inhabitants, and human populations. Any array of contiguous soils influenced by local flora and fauna is termed a biosequence. The vegetation observed along this narrow foothill region varies from shrubs in the dry south to needle-leaved trees in the humid north, with extensive grasslands in between. In the middle of the precipitation range, transition zones occur in which small groves of needle-leaved trees are interspersed with grassland patches in an apparently random manner. These plant populations represent local flora largely selected by climate. The properties of the soils underlying these plants, however, exhibit differences that do not arise from climate, topography, or parent material but are an effect of the differing plant species. The soils under trees, for instance, are much more acidic and contain much less humus than those under grass, and nitrogen content is considerably greater in the grassland soil. These properties come directly from the type of litter produced by the two different kinds of vegetation.

An opportunity to examine biosequences is often presented by relatively young soils formed from an alluvial parent material. Soils of this kind lying beneath shrubs may be richer in humus and plant nutrients than similar soils found beneath needle-leaved trees. This variation results from differences in the cyclic processes of plant growth, litter production, and litter decay. Organic matter decomposers will feed on stored material in soil if litter production is low, whereas high litter production will permit soil stocks of organic matter to increase, leading to humus-rich A horizons as opposed to the leached E horizons found in soils that form under humid climatic conditions.

Human beings are also part of the biological influx that influences soil formation. Human influence can be as severe as wholesale removal or burial (by urbanization) of an entire soil profile, or it can be as subtle as a gradual modification of organic matter by agriculture or of soil structure by irrigation. The chemical and physical properties of soils critical to the growth of crops often are affected significantly by cultural practices. Among the problems created for agriculture by cultural practices themselves are loss of arable land, erosion, the buildup of salinity, and the depletion of organic matter.

Anthrosol soil profile from the Netherlands showing surface layers made homogeneous through human activity.

Time

The soil-forming factors of parent material and topography are largely site-related (attributes of the terrain), whereas those of climate and organisms are largely flux-related (inputs from the surroundings). Time as a soil-forming factor is neither a property of the terrain nor a source of external stimulus. It is instead an abstract variable whose significance is solely as a marker of the evolution of soil characteristics. The conceptual independence of time from its four companion factors means simply that soil evolution can occur while site attributes and external inputs remain essentially unchanged.

Certain soil profile features can be interpreted as indicators of the passage of time. (A series of soil profiles whose features differ only as a result of age constitutes a chronosequence.) One example of a time-related feature is the humus content of the A horizon, which, for soils less than 10,000 years old, increases continually at a rate dependent on parent material, vegetation, and climate. Typically, this rate of increase slows after about 10,000 years, plant nutrients begin to leach away, and a significant decline in humus content is observed for soils whose age approaches one million years. (Agricultural practices can interrupt this trend, causing a gradual drop in stored humus by 25 percent or more. Correlated with these changes are economically important soil properties, such as nutrient supply and retention capability, acidity, and aeration.)

The accumulation of clay and lime in soil profiles as a result of their translocation downward is also an indication of aging. For example, older soils that have formed on calcium-containing loess deposits have better-developed E and Bt horizons (as well as thinner A horizons) than younger soils forming on these deposits. Similarly, soils in a chronosequence developed on alluvium can exhibit a clayey hardpan after 100,000 years or so. Soils also tend to redden in colour as they age, irrespective of climatic conditions, reflecting the persistence of poorly crystalline or crystalline oxide minerals containing Fe^{3+}. Indeed, the dominant mineralogy of the clay-size particles in soils is itself a reliable indicator of soil age. Any particular sequence of predominant clay mineralogy found in a soil is known collectively as the set of Jackson-Sherman weathering stages. Each downward increment through the table corresponds to increasing mineral residence time, both among and within the three principal stages (early, intermediate, and advanced).

Jackson-Sherman soil weathering stages			
	Characteristic minerals in soil clay fraction	Characteristic chemical and physical conditions of soil	Characteristic soil profile features
Early stage	Gypsum, carbonates, olivine/pyroxene/Amphibole, Fe^{2+}-bearing micas, and feldspars	Low water and humus content, limited leaching, reducing environments, and limited time for weathering	Minimally weathered soils all over the world, though mainly in arid regions where low rainfall keeps weathering to a minimum
Intermediate stage	Quartz, mica/illite, vermiculite/chlorite, and smectite	Retention of Na, K, Ca, Mg, Fe^{2+}, and silica; alkalinity and ineffective leaching; igneous rock rich in Ca, Mg, and Fe^{2+} but no Fe^{2+} oxides; easily hydrolyzed silicates; and transport of silica into the weathering zone	Soils of temperate regions developed under grass or trees—i.e., the major agricultural soils of the world
Advanced stage	Kaolin, aluminum and iron oxides, and titanium oxides	Removal of Na, K, Ca, Mg, Fe^{2+}, and silica; effective leaching by fresh water; low pH; and dispersion of silica	Intensely weathered soils of the humid tropics, frequently characterized by acidity and low fertility

The early stage of weathering is recognized through the dominance of sulfates, carbonates, and primary silicates, other than quartz and muscovite, in the soil clay fraction. These minerals can survive only if soils remain very dry, very cold, or very wet most of the time—that is, if they have limited exposure to water, air, or solar energy. The intermediate stage features quartz, muscovite, and secondary aluminosilicates prominently in the clay fraction. These minerals can survive under leaching conditions that do not deplete silica and metals, such as calcium or magnesium, and that do not result in the complete oxidation of Fe^{2+}, which is then incorporated into illite and smectite clays. The advanced stage, on the other hand, is associated with intensive leaching and strong oxidizing conditions, such that only oxides of aluminum, Fe^{3+}, and titanium remain. Kaolin will be a dominant clay mineral group only if the removal of silica by leaching is not complete or if there is an encroachment of silica-rich waters—as can occur, for example, when water percolating through a soil profile at the upper part of a toposequence moves downslope into a soil profile at the lower part.

Implicit is the conclusion that more time is required to form soils featuring the more persistent minerals in the clay fraction. This conclusion is borne out by careful field studies worldwide in which the rate of soil horizon formation is determined. Soils whose clay mineralogy falls in the early stage require less than a decade to develop a centimetre (0.4 inch) of horizon thickness. Soils with dominant clay minerals in the intermediate stage do this in less than a century, whereas soils with dominant clay minerals in the advanced stage need several hundred years to form a centimetre of solum.

Weathering

The disintegration and decomposition of rocks and rock minerals by physical, chemical and biological agents in single or in combination is termed as weathering. Two types of weathering are recognized: (1) Geochemical weathering occurs before soil is developed and continues beneath soil cover. It is also termed as "geogenesis" and (2) Pedochemical weathering that occurs as result of disintegration and transformation of minerals that take place within the soil. It is also termed as "pedogenesis". Weathering on the basis of agents caused can be classified into following three types. The Physical/ Mechanical weathering takes place when rock is split or broken into smaller pieces of the same material without changing its composition. Agents include temperature, water, ice, gravity, wind, mutual rubbing of rock fragments and air blast of sand grains. It is more predominant in the desert and arctic regions. Biological weathering occurs due to weakening and subsequent disintegration of rock by plants, animal and microbes. Their agents include micro-organisms such as bacteria, fungi, protozoa, lichens and mosses and also burrowing animals like earthworms. These micro- organisms decompose a part of the organic matter and add inorganic nutrients to soil. Chemical weathering includes processes whereby rock materials are decomposed by alteration of present material. Their agents include atmospheric gases, especially oxygen, water and acids. It takes place through six different chemical reactions:

- Solution: Weathering of water soluble minerals (limestone, gypsum) by solvent action. It increases in presence of carbon dioxide and organic acids formed as a result of decay of organic plant and animal remains.
- Oxidation: Oxides dissolve in water weakening rock and causing weathering. It occurs best in well aerated and well drained soils.

$$4Fe + 3O_2 \rightarrow 2Fe_2O_3$$

- Hydrolysis: Here, weathering by the chemical reaction occurs between hydrogen ions and hydroxyl ions of water and a mineral's ions.

$$2KAlSi_3O_8 + 2H^+ + 2HCO_3^- + H_2O \rightarrow Al_2Si_2O_5(OH)_4 + 4SiO_2 + +2K^+ + 2HCO_3$$

- Reduction: It occurs in poorly aerated deep zones of earth's crust. The red ferric oxide reduces to grey ferrous oxide.

$$2Fe_2O_3 \rightarrow 4Fe + O_2$$

- Carbonation: Combination of carbon dioxide and water to form carbonic acid. Carbonic acid combines with hydroxides of Ca, Mg to form carbonates and bicarbonates.
- Hydration: Attack of water molecule to the rock material resulting in increment in volume and softening of hydrated mineral with its subsequent weathering.

$2Fe_2O_3 + 3H_2O \rightarrow 2Fe_2O_3.3H_2$

Physical Weathering

Sediment comes from the breakdown of rocks into smaller, transportable components. This occurs via two processes: physical weathering and chemical weathering. Physical weathering consists of breaking apart rocks and crystals through different processes without changing their chemical composition. The results of physical weathering are smaller components of the same material that is being weathered. There is no change in chemical composition. Physical weathering tends to produce mostly sand-sized sediment and larger grains because most of the fracturing occurs along mineral boundaries. Physical weathering of fine grained or finely crystalline rock can produce abundant very fine grains, but most of the sediment from these rock types consists of rock fragments called lithic clasts. Lithic clasts produced from physical weathering range in size from very fine silts and clays to large boulders and gravel.

Mountain regions undergo rapid physical weathering due to wind, ice, and temperature changes.

Physical weathering, also known as mechanical weathering, works along with chemical weathering to efficiently wear down rocks. Breaking a rock through physical weathering increases the effective surface area in which chemical weathering can occur, and through changing the composition of rocks in chemical weathering, the durability of a rocks surface can decrease allowing for it to be more easily removed by physical processes. From this, we can see that both chemical and physical weathering work together to increase rock erodibilty.

Physical Weathering Processes

There are 6 common ways in which physical weathering happens:

Abrasion

Abrasion is the process by which clasts are broken through direct collisions with other clasts. Gravity causes abrasion as rocks tumble down a mountainside or cliff. Abrasion also commonly occurs in river beds where clasts tumble along the bottom with the current or on sand dunes where wind causes grains of sand and silt to collide with exposed rock. Rocks can also undergo abrasion within glaciers too, where clasts embedded in ice grind along the rock below. Clasts that undergo abrasion tend to become more round and smooth. This process is called rounding.

Abraded rocks along river stream bed. Note the rounded and flattened shape.

Frost Wedging

Frost wedging occurs in places that have the right temperatures to freeze water and then melt water. This commonly happens in Polar Regions and mid latitude mountains where sunlight can melt water during the day and refreeze overnight when temperatures drop. Frost wedging is directly related to the fact that as water freezes its volume increases by roughly 9%. Liquid water percolates into existing cracks within a rock. The pressure of expanding ice causes cracks to widen and extend. This processes then repeats as melted water further fills newly formed cracks and freezes. Frost wedging works quickly and results of which can be seen at the bottom of hill-slopes as piles of fragmented rocks.

Biological Activity/Root Wedging

Weathering processes can happen due to the activity of living organisms. Burrowing animals can break rocks and stir sediments causing physical weathering. Animals that burrow can also bring fresh material to the surface where physical and chemical weathering can work more efficiently. Plant roots in search of nutrients in water grow into fractures. As the roots grow they wedge the rock apart similar to the frost wedging process. This is called root wedging. During root growth, organic acids can form contributing to chemical weathering.

Salt Crystal Growth

Another expansive force that can break rocks is created by the growth of salt crystals. This process frequently happens near shorelines and in arid regions ie: beaches and deserts. The process begins when water that has dissolved salt in it penetrates cracks in rocks. As the water evaporates, salt crystal begins to form. These crystals expand pushing against existing grains in the rock, this eventually breaks it.

Sheeting

Sheeting, also known as exfoliation, happens in regions with large masses of igneous rocks. It is caused by the pressure release as overburden rock is eroded away. The process of removing overlying rock is called unloading. As the pressure is released, expansion of the rock causes concentric layers of cracks to form within the igneous body. These layered "sheets" are then broken off by continued weathering.

Half Dome is commonly used as an example of exfoliation. Note the large fractured sheets.

Thermal Expansion

When minerals are subjected to different temperature ranges they expand and contract. Rapid temperature fluctuations such as day night cycles can cause individual grains in rocks to expand and contract at different rates. Individual grains can be squeezed from the rock surface or fractures can form to relieve stress. The efficiency of this method is debated by geologist, but no less contributes to physical weathering processes. Climates that experience rapid temperature changes such as deserts display this weathering.

Chemical Weathering

Minerals weather chemically when they react with air and water. In these processes some minerals dissolve and others combine with atmospheric components (mainly water, oxygen and carbon dioxide) to form new minerals. The specific reactions include solution, oxidation, hydration and hydrolysis.

Profile of weathered sedimentary rocks to limonites and clay minerals.

Profile of metamorphic rocks weathered to clay minerals.

Carbon dioxide (CO_2) released from decaying organic matter and from the atmosphere dissolved in rainwater form the weak carbonic acid:

$$CO_2 + H_2O \rightarrow H_2CO_3 \rightarrow H^+ + (HCO_3)^-$$

This acid attacks limestone, dissolving carbonates (e.g., $CaCO_3$) yielding an aqueous solution of calcium and bicarbonate ions:

$$\underset{\text{Limestone}}{CaCO_3} + \underset{\text{Carbonic acid}}{H_2CO_3} \rightarrow \underset{\text{Calcium ions}}{Ca^{++}} + \underset{\text{Bicarbonate ions}}{2(HCO_3)^-}$$

Dissolution of limestone may create extensive underground cavern systems.

Oxidation and Hydration: Oxidation produces iron oxide minerals (hematite and limonite) in well aerated soils, usually in the presence of water. Pyroxene, amphibole, magnetite, pyrite, and olivine are most susceptible to oxidation because they have high iron content. Iron contained in minerals combines with oxygen and water to form hydrated iron oxides as follows:

$$\underset{\text{Iron-rich minerals}}{4Fe^{++}} + \underset{\text{Oxygen}}{3O_2} + \underset{\text{Water}}{6H_2O} \rightarrow \underset{\text{Limonite}}{2(Fe_2O_3 3H_2O)}$$

Hydrolysis: This reaction is responsible for the formation of clays, the most important mineral in soils. A typical hydrolytic reaction occurs when orthoclase feldspar reacts with slightly acidic water to form clay minerals, potassium ions, and silica in solution:

$$\underset{\text{Orthoclase feldspar}}{2KAlSi_3O_8} + \underset{\text{acid}}{2H^+} + \underset{\text{water}}{9H_2O} \rightarrow \underset{\text{clay mineral}}{Al2Si_2O_5(OH)_4} + \underset{\text{potassium ion}}{2K^+} + \underset{\text{soluble silica}}{4H_4SiO_4}$$

The ions released from silicate minerals in the weathering process are sodium, potassium, calcium, iron, and magnesium ions. They are carried away by rain and river waters or become important soil nutrients.

Relative Stability of Common Minerals Under Weathering	
Stability of minerals	**Rate of weathering**
MOST STABLE	Slowest
Iron oxide (hematite)	
Aluminum hydroxides (gibbsite)	
Quartz	
Clay minerals	
Muscovite mica	
Potassium feldspar	
Biotite	
Albite (Na-rich feldspar)	
Amphiboles	
Pyroxene	
Anorthite (Ca-rich feldspar)	
Olivine	
Calcite	
Halite	
LEAST STABLE	Fastest

The rate of chemical weathering is controlled by:

- The chemical stability of minerals: This is related to the tendency of a mineral to remain in a given chemical form rather than to react spontaneously to become a different chemical substance. The stability of a specific mineral is determined by the specific set of conditions (e.g. pressure and temperature) under which they formed. Quartz, for example, is the last mineral to crystallize from a silicate magma and has a greater chemical stability under physical conditions at the Earth's surface than olivine, pyroxene, and feldspar. These minerals crystallize earlier and at higher temperatures and pressure, are therefore least stable and most subject to chemical weathering. Relative stability of various minerals enables us to ascertain the intensity of weathering in a given area.
- Surface area: A massive volume of unfractured rock is less susceptible to weathering than an equal mass of fractured rock. This is because the massive rock has less surface area on which atmospheric processes can act, whereas smaller fragments have a larger area upon which chemical and physical weathering can act.
- Climate: Chemical weathering dominates in humid areas.

Biological Weathering

Biological weathering only refers to weathering caused by organisms -- animals, plants, fungi and microorganisms such as bacteria. While certain forms of biological weathering, such as the breaking of rock by tree roots, are sometimes categorized as either physical or chemical, biological weathering can be either physical or chemical. Biological weathering can work hand in hand with physical weathering by weakening rock or exposing it to the forces of physical or chemical weathering.

Trees and Other Plants

You've probably seen biological weathering in action if you've ever seen a sidewalk that's cracked by a tree root. The roots of trees, grasses and other plants can grow into small spaces and gaps in rock. When these roots grow, they exert pressure on the rock around them, causing the gaps to widen or even crack. Plant roots can also weather rock through chemical processes. When dead roots decompose, they release carbon dioxide; this is sometimes converted into carbonic acid, which chemically breaks down rock into soil.

Microorganisms and Lichens

Not all biological weathering occurs visibly. Many microorganisms in the soil and on the surface of rock can contribute. Some bacteria derive nutrition by taking a combination of nitrogen from the air and minerals -- such as silica, phosphorous and calcium -- from rock. By removing these minerals, the rock is weakened and is further subject to other weathering forces such as wind and water. Lichens, symbiotic colonies of fungi and microscopic algae that grow on rock, also contribute to weathering. The fungi in lichen produce chemicals that break down the minerals in the rock. The algae, like the bacteria, use these minerals for nutrition.

Animal Activity

Animals can also contribute to weathering. Animals can walk on rock or disturb it, causing landslides that scrape or smooth rock surfaces. Burrowing animals such as badgers and moles can break up rock underground or bring it to the surface, where it is exposed to other weathering forces. Some animals directly burrow into the rock. The piddock shell is a mollusk, closely related to the clam that uses its shell to cut a hole in rock, where it lives.

As animals, humans also contribute to biological weathering. Construction, mining and quarrying break up and disturb large sections of rock. Foot traffic over rock causes friction that breaks off tiny particles. Over a long period, foot traffic can cause significant wear and tear on rock surfaces.

Soil Forming Factors

Soil forms continuously, but slowly, from the gradual breakdown of rocks through weathering. Weathering can be a physical, chemical or biological process:

- Physical weathering: Breakdown of rocks from the result of a mechanical action. Temperature changes, abrasion (when rocks collide with each other) or frost can all cause rocks to break down.
- Chemical weathering: Breakdown of rocks through a change in their chemical makeup. This can happen when the minerals within rocks react with water, air or other chemicals.
- Biological weathering: The breakdown of rocks by living things. Burrowing animals help water and air get into rock, and plant roots can grow into cracks in the rock, making it split.

The accumulation of material through the action of water, wind and gravity also contributes to soil formation. These processes can be very slow, taking many tens of thousands of years. Five main interacting factors affect the formation of soil:

- Parent material: Minerals forming the basis of soil.
- Living organisms: Influencing soil formation.
- Climate: Affecting the rate of weathering and organic decomposition.
- Topography: Grade of slope affecting drainage, erosion and deposition.
- Time: Influencing soil properties.

Interactions between these factors produce an infinite variety of soils across the earth's surface.

Soil Composition

Soil composition is an important aspect of nutrient management. The soil consists of a mixture of solid particles, water and air. Solid particles can be inorganic or organic. The inorganic parts are mainly minerals: silicates, oxides and hydroxides of iron, aluminium, manganese, etc. that, according to their dimension, are classified in skeleton and fine earth, which are then divided into sand, slime and clay. These derive from the alteration of rocks into smaller and incoherent material, which accumulates to form superficial deposits. The deposit of incoherent material can occur in the same place as the rock was disintegrated.

The degradation processes of vegetal residues (leaves, fruits, dry branches or whole plants) and dead animals lead to the creation of organic fractions of the soil. Organic compounds can keep unaltered for long periods (non-humic compounds) or be subject to deep and fast changes in their original chemical structure (humic compounds or humus).

Water and air occupy the free spaces between solid particles (pores), and form a thick and extended network that allows water to move in the ground.

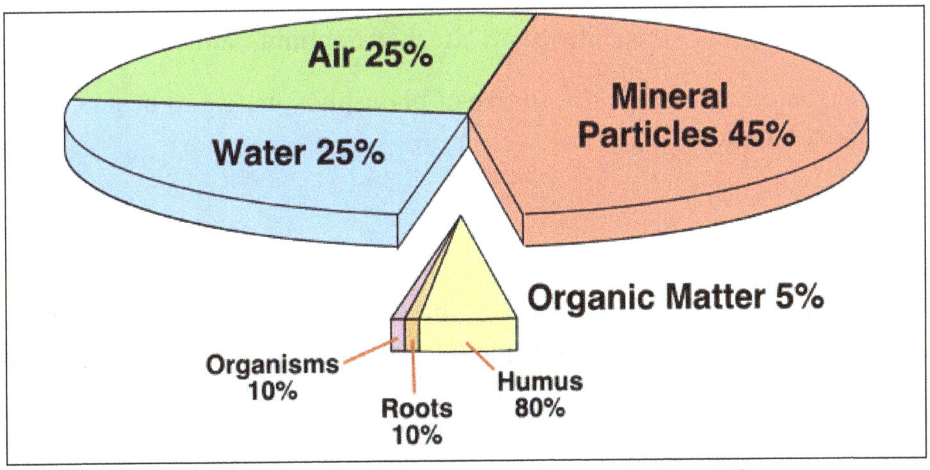

Constituents of soil (by percentage) in a typical top soil.

- Mineral matter: Geochemically classified on basis of their bonding characteristics into five main groups:

Table: Components of Mineral matter of soil

Element	Characteristics	Examples
Lithophile	Readily ionized forming oxyanions.	Li, Na, K, Ca, Mg, Fe, Si, B, Bi, Ni, Co, S, Cu, Zn, halogens, rare earths.
Chalcophile	Forms covalent bonds with sulphides.	Fe, Ni, Co, Cu, Zn, Pb, Ag, Sb, As, Bi, S, Se, Te, As.
Siderophile	Capable of forming metallic bonds.	Fe, Co, Pd, Os, Au.
Atmosphile	Tend to remain in atmospheric gases.	He, Ne, Ar, N, O
Biophile	Remain associated with living organisms.	C, H, O, N, P, S, Ca, Mg

- Organic matter: It varies from 3.5 – 5% by weight in top soil. It contains partly decayed plant and animal residues that are microbially synthesized contributing in maintenance and improvement of soil properties. It acts as storehouse of plant nutrients. Five major classes of organic matter exist in soil namely humus, saccharides, fats, resins and waxes, nitrogen organics and phosphorus containing organics:
    - Humus: The formation by microbial degradation of plants and animals is called humification. The composition of humic substances follows order: C: 45-55%,O :30-45%, H :3-6%, N :1- 5%, S :1%. It possesses aromatic character and are instrumental in decomposing soil minerals.
    - Saccharides: It assist in stabilization of soil aggregates from major food sources. e.g. sugar, cellulose, starches, gums.
    - Fats, resins and waxes: They are lipid extractable. The enzyme lipase degrades fats into glycerol and fatty acids which further breaks down into carbon dioxide and water. Being phototoxic affects soil properties by repelling water.
    - Nitrogen organics: In this, nitrogen is attached to humus, amino acids and amino sugars.
    - Phosphorus organics: It is the source of plant phosphate and occurs as phosphate esters, phospholipids.

Soil Water

It provides medium to plants for obtaining necessary nutrients. Itaids in maintenance of texture, arrangement and compactness of soil particles. The total amount of water present in the soil is called holard. Water that can be absorbed by plants ischesard while unabsorbed is called echard.

Soil Air

Its composition nearly similar to atmospheric air except CO_2 level being high. Water clogging between air spaces leads to anaerobic condition in which gases are unable to diffuse inwards. It is essential for respiration of soil microorganisms and plants and enhances permeability of roots to soil.

Organic and Inorganic Soil Components

Soil is a mixture of organic and inorganic constituents. The inorganic components come from the weathering of various rock types. Organic particles are the result of both plants and animals inhabiting the area. While some soil has only one particle type, the most fertile soil is a mixture of several organic and inorganic components.

Inorganic Components of Soil

Most of the inorganic components of soil exist as silicate minerals. The common elements in the soil are oxygen (46.6 per cent), silicon (27.7 per cent), aluminium (8.1 per cent), iron (5.0 per cent), calcium (3.6 per cent), sodium (2.8 per cent), potassium (2.6 per cent) and magnesium (2.1 per cent). Commonly found silicates in the soil are quartz (SiO_2), Orthoclase ($KAlSi_{308}$); albite ($NaAlSi_3O_8$) and epidote [$4CaO.3(AlFe)_2O_3 6SiO_2H_2O$]. Many soils contain iron oxides, FeO (OH), magnetite (Fe_3,O_4). Certain soils contain titanium, and manganese oxides. Calcium carbonate is another common component of soil.

Soil also contains a mixture of minerals as carbonates, sulphates, chlorides and nitrates. Some soils also contain clay mineral, which are hydrated iron and aluminum silicates. These clay minerals bind cations such as Ca^{2+}, Mg^{2+}, Na^+, K^+, NH_4^+ so that these cations are not leached by water and are available to plants as plant nutrients.

Organic Components of Soil

Soil organic matter (SOM) is the organic matter component of soil. Soil organic matter is the fraction of the soil that consists of plant or animal tissue in various stages of breakdown (decomposition). Most of our productive agricultural soils have between 3 and 6% organic matter.

Soil organic matter contributes to soil productivity in many different ways. In this fact sheet, we describe the various components of organic matter and the different roles organic matter plays in soil productivity. We also discuss field management practices that will help preserve or increase soil organic matter levels over time. Organic matter is made up of different components that can be grouped into three major types:

- Plant residues and living microbial biomass.
- Active soil organic matter also referred to as detritus.
- Stable soil organic matter, often referred to as humus.

The living microbial biomass includes the microorganisms responsible for decomposition (breakdown) of both plant residues and active soil organic matter or detritus. Humus is the stable fraction of the soil organic matter that is formed from decomposed plant and animal tissue. It is the final product of decomposition.

The first two types of organic matter contribute to soil fertility because the breakdown of these fractions results in the release of plant nutrients such as nitrogen, phosphorus, potassium, etc.

The humus fraction has less influence on soil fertility because it is the final product of decomposition (hence the term "stable organic matter"). However, it is still important for soil fertility management because it contributes to soil structure, soil tilth, and cation exchange capacity. This is also the fraction that darkens the soil's color.

Benefits of Stable Soil Organic Matter

There are numerous benefits to having a relatively high stable organic matter level in an agricultural soil. These benefits can be grouped into three categories:

Physical Benefits:

- Enhances aggregate stability, improving water infiltration and soil aeration, reducing runoff.
- Improves water holding capacity.
- Reduces the stickiness of clay soils making them easier to till.
- Reduces surface crusting, facilitating seedbed preparation.

Chemical Benefits:

- Increases the soil's CEC or its ability to hold onto and supply over time essential nutrients such as calcium, magnesium and potassium.
- Improves the ability of a soil to resist pH change; this is also known as buffering capacity.
- Accelerates decomposition of soil minerals over time, making the nutrients in the minerals available for plant uptake.

Biological Benefits:

- Provides food for the living organisms in the soil.
- Enhances soil microbial biodiversity and activity which can help in the suppression of diseases and pests.
- Enhances pore space through the actions of soil microorganisms. This helps to increase infiltration and reduce runoff.

Organic Materials

Over time, the application and incorporation of organic materials can result in an increase in stable soil organic matter levels. Sources of organic materials include:

- Crop residues
- Animal manure
- Compost

- Cover crops (green manure)
- Perennial grasses and legumes

The quickest increases are obtained with sources that are high in carbon such as compost or semi-solid manure.

Compost application can increase soil organic matter levels over time.

Organic Matter Management

Farm practices that help to maintain or increase soil organic matter levels:

- Use of conservation tillage practices (for example zone tillage or no-till). Tillage exposes the organic matter to air and will result in the lowering of stable organic matter due to increased mineralization rates and erosion losses.
- Rotation of annual row crops with perennial grass or legume sods will reduce erosion and build up organic matter as a result of the decomposition of the rootmass.
- Establishment of legume cover crops will enhance organic matter accumulation by providing the nitrogen (N) needed for decomposition of freshly added organic materials, especially those with a high C to N ratio (corn stover, cereal straw, heavily bedded manure, etc).
- Avoiding soil compaction which increases waterlogging, and maintaining proper pH to enhance microbial activity and decomposition of freshly added materials.

Actual build-up of stable organic matter will, in addition to the amount and source of organic material added, and tillage and rotation practices, also depend on:

- Soil temperature
- Precipitation and soil moisture holding capacity
- Soil type and drainage class
- Existing microbial community
- Soil fertility status and soil pH

Soil Structure

The texture and structure of soil influences the amount of water held and hence behavior of any soil. Soil bulk density with soil texture influences water infiltration rate. Clay soil has slow water drainage than sand and volcanic soils. A loam soil with humus containing 34% of air and 66% of water is considered best for majority of crops. Secondary particles called aggregates or peds formed by the arrangement or grouping of primary particles (sand, clay, slit and organic matter) is defined as soil structure.

Formation of Soil Structure

The process of soil formation occurs through physio-chemical process characterized by aggregate formation, flocculation and swelling, shrinking of clay masses and biological process where alleviation of aggregates by burrowing animals, binding action of plant roots and production of organic glues by micro-organisms occurs. Soil is known to contain air spaces and generally possess a loose texture.

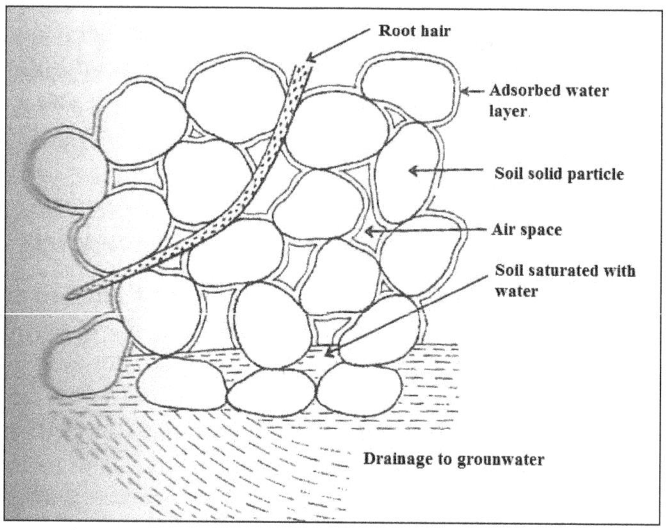

Fine structure of soil showing solid, water, air phases.

The arrangement of soil aggregates into different forms gives a soil its structure. The natural processes that aid in forming aggregates are:

- Wetting and drying.
- Freezing and thawing.
- Microbial activity that aids in the decay of organic matter.
- Activity of roots and soil animals.
- Adsorbed cations.

The wetting/drying and freezing/thawing action as well as root or animal activity pushes particles back and forth to form aggregates. Decaying plant residues and microbial byproducts coat soil particles and bind particles into aggregates. Adsorbed cations help form aggregates whenever a cation is bonded to two or more particles. Aggregates are described by their shape, size and stability.

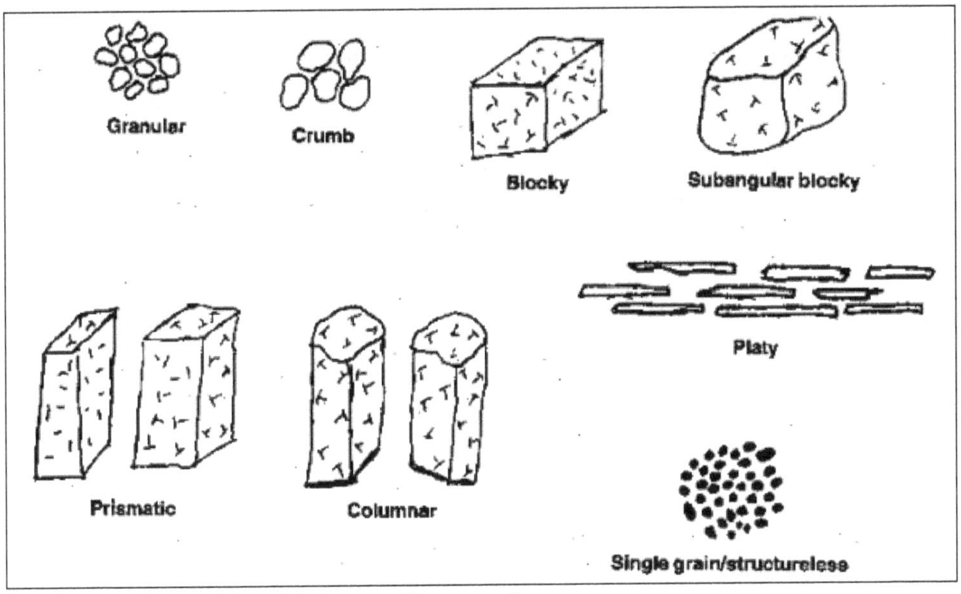

Soil structural types.

Types of Soil Structure

There are five major classes of structure seen in soils: platy, prismatic, columnar, granular, and blocky. There is also structure less conditions. Some soils have simple structure, each unit being an entity without component smaller units. Others have compound structure, in which large units are composed of smaller units separated by persistent planes of weakness.

- Platy: In platy structure, the units are flat and plate-like. They are generally oriented horizontally. A special form, lenticular platy structure, is recognized for plates that are thickest in the middle and thin toward the edges. Platy structure is usually found in subsurface soils that have been subject to leaching or compaction by animals or machinery. The plates can be separated with a little effort by prying the horizontal layers with a pen knife. Platy structure tends to impede the downward movement of water and plant roots through the soil.
- Prismatic: In prismatic structure, the individual units are bounded by flat to rounded vertical faces. Units are distinctly longer vertically, and the faces are typically casts or molds of adjoining units. Vertices are angular or sub-rounded; the tops of the prisms are somewhat indistinct and normally flat. Prismatic structures are characteristic of the B horizons or sub-soils. The vertical cracks result from freezing and thawing and wetting and drying as well as the downward movement of water and roots.
- Columnar: In columnar structure, the units are similar to prisms and are bounded by flat or slightly rounded vertical faces. The tops of columns, in contrast to those of prisms, are very distinct and normally rounded. Columnar structure is common in the subsoil of sodium affected soils. Columnar structure is very dense and it is very difficult for plant roots to penetrate these layers. Techniques such as deep plowing have help to restore some degree of fertility to these soils.
- Blocky: In blocky structure, the structural units are block-like or polyhedral. They are

bounded by flat or slightly rounded surfaces that are casts of the faces of surrounding peds. Typically, blocky structural units are nearly equi-dimensional but grade to prisms and to plates. The structure is described as angular blocky if the faces intersect at relatively sharp angles; as sub-angular blocky if the faces are a mixture of rounded and plane faces and the corners are mostly rounded. Blocky structures are common in subsoil but also occur in surface soils that have a high clay content. The strongest blocky structure is formed as a result of swelling and shrinking of the clay minerals which produce cracks. Sometimes the surface of dried-up sloughs and ponds shows characteristic cracking and peeling due to clays.

- Granular: In granular structure, the structural units are approximately spherical or polyhedral and are bounded by curved or very irregular faces that are not casts of adjoining peds. In other words, they look like cookie crumbs. Granular structure is common in the surface soils of rich grasslands and highly amended garden soils with high organic matter content. Soil mineral particles are both separated and bridged by organic matter breakdown products, and soil biota exudates, making the soil easy to work. Cultivation, earthworms, frost action and rodents mix the soil and decrease the size of the peds. This structure allows for good porosity and easy movement of air and water. This combination of ease in tillage, good moisture and air handling capabilities, and good structure for planting and germination, are definitive of the phrase good tilth.
- Structureless: Some soils lack structure and are referred to as structureless. In structureless layers or horizons, no unit are observable in place or after the soil has been gently disturbed, such as by tapping a spade containing a slice of soil against a hard surface or dropping a large fragment on the ground. When structureless soils are ruptured, soil fragments, single grains, or both result. Structureless soil material may be either single grain or massive. Soil material of single grains lacks structure. In addition, it is loose.

Factors Affecting Soil Structure

The development of structure in arable soil depends on the following factors:

- Climate: Climate has considerable influence on the degree of aggregation as well as o11 the type of structure. In arid region, there is very little aggregation of primary particles. In semi- arid regions, the degree of aggregation is greater than arid regions.
- Organic matter: Organic matter improves the structure of a sandy soil as well as of a clay soil. In a case of sandy soil, the sticky and slimy material produced by the decomposing organic matter and the associated microorganism cement the sand particles to form aggregates. In the case of clayey soil, it modifies the properties of clay by reducing its cohesive power. This helps making clay more crumby.
- Tillage: Cultivation implements break down of large clods into smaller fragments and aggregates. For obtaining good granular and crumby structure, optimum moisture content in the soil is necessary. If the moisture content is too high it will form large clods on drying. If it is too low, some of the existing aggregates will be broken down.

- Plant roots: Large numbers of granules remain attached to roots and root hairs which help to develop crumb structure. Plant root secretions may also act as cementing agents in binding the soil particles. The plant roots, on decay, may also bring about granulation due to the production of sticky substances.
- Soil organism: Among the soil fauna, small animals like earthworms, moles and insects etc., that burrow in the soil are the chief agents that take part in the aggregation of finer particles.
- Fertilizers: Fertilizer like Sodium nitrate destroys granulation by reducing the stability of aggregates. Few fertilizers, for example, Calcium Ammonium nitrate, help in development of good structures.
- Wetting and drying: Wren a dry soil is wetted, the soil colloids swell on absorbing water. On drying, shrinkage produced strains in the soil mass give rise to cracks which break it up into clods and granules of various sizes.

Effects of Soil Structure on Other Physical Properties of Soil

Soil structure brings change in other physical properties of soil like porosity, temperature, density, consistency and colour.

- Porosity: Porosity of a soil is easily changed. In plate-like structure pore spaces are less whereas in crumby structure pore spaces are more.
- Temperature: Crumby structure provides good aeration and percolation in the soil. Thus, these characteristics help in keeping optimum temperature in comparison to plate-like structure.
- Density: Bulk density varies with the total pore space present in the soil. Structure chiefly influences pore spaces. Platy structure with less total pore spaces has high bulk density whereas crumby structure with more total pore spaces has low bulk density.
- Consistence: Consistence of soil also depends on structure. Plate-like structure exhibits strong plasticity.
- Colour: Bluish and greenish colours of soil are generally due to poor drainage of soil. Platy structure normally hinders free drainage.

Structural Management of Soils

Coarse-Textured Soil

Sandy soils are commonly too loose and lack the capacity to adsorb and hold sufficient moisture and nutrients. They lack fertility and water- holding capacity. There is only one practical method of improving the structure of such soil- the addition of organic matter. Organic matter will not only act as a binding agent for the particles but will also increase the water-holding capacity. Sod-crops, for example, corn, blue grass etc., also help in improving the structural condition of sandy soils.

Fine-Textured Soil

The structural management of a clay soil is difficult than sandy soil. In clay, plasticity and cohesion are high because of the presence of large amount of colloidal clay. When such a soil is tilled when wet, its pore space becomes much reduced, it becomes practically impervious to air and water and it is said to be puddled. When a soil in this condition dries, it usually becomes hard and dense.

The tillage of clay soil should be done at right moisture stage. If ploughed too wet, the structural aggregates are broken down and an un-favourable structure results. On the other hand, if ploughed too dry, big clods are turned up which are difficult to work. The granulation of fine-textured soil should be encouraged by the incorporation of organic matter. Growing of sod-crops also improves granulation in the soil.

Rice Soil

Puddling of the soil is generally beneficial to the production of rice. In preparation for the planting of rice, the soil is flooded with water and then puddled by intensive tillage. Puddling destroys the structural aggregates. Rice seedling is transplanted into the freshly prepared mud.

Such soil management helps control weeds and also reduce the rate of water movement down (percolation) through the soil. This is important to maintain standing water in the rice throughout the growing season. By reducing water percolation, puddled soil markedly decreases the amount of water needed to produce a rice crop.

Semi-aquatic characteristics of the rice plant account for its positive response to a type of soil management that destroy aggregate. Rice survives flooded conditions because oxygen moves downward inside the stem of the plant to supply the roots. This characteristic permits rice to stand well in the water-logging condition. Rice can be grown successfully on un-puddled but flooded soil.

Soil Grade

It describes the uniqueness of the structure that is further combined with the cohesion of soil within units compared to cohesion between individual units. Weak Grade is characterized by poor observation of aggregates where peds break into whole and broken units on disturbance. In the moderate type of grade, the structural units are easily identified. Peds break into mixture of entire, broken & little aggregates, on shaking. The 'clean' separation from other peds is noticeable. Strong grade is characterized by distinct structural units and identifiable shape. Peds retain their shape on shaking (clean separation).

Structural Class

It describes the size of units. The smallest dimension of any structural unit is referred by its size limit.

Table: Structural classes of soil.

Shapes of structures (mm)				
Size classes	Platy	Prismatic & Columnar	Blocky	Granular
Very fine	< 1	< 10	< 5	< 1
Fine	1 – 2	10 – 20	5 – 10	1 – 2

Medium	2 – 5	20 – 50	10 – 20	2 – 5
Coarse	5 – 10	50 – 100	20 – 50	5 – 10
Very coarse	> 10	> 100	> 50	> 10

Soil Profile

Soil profile is defined as the vertical section of the soil from the ground surface downwards to where the soil meets the underlying rock.

The soil is the topmost layer of the earth's crust mainly composed of organic minerals and rock particles that support life. A soil profile is a vertical cross-section of the soil, made of layers running parallel to the surface. These layers are known as soil horizons.

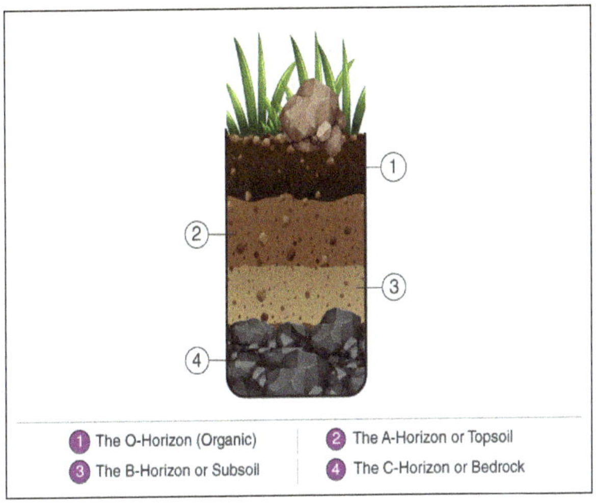

The soil is arranged in layers or horizons during its formation. These layers or horizons are known as the soil profile. It is the vertical section of the soil that is exposed by a soil pit. The layers of soil can easily be identified by the soil colour and size of soil particles. The different layers of soil are:

- Topsoil
- Subsoil
- Parent rock

Each layer of soil has distinct characteristics. Soil profile helps in determining the role of the soil as well. It helps one to differentiate the given sample of soil from other soil samples based on factors like its colour, texture, structure, and thickness, as well as its chemical composition.

Layers of Soil

The soil profile is composed of a series of horizons or layers of soil stacked one on top of the other. These layers or horizons are represented by letters O, A, E, C, B and R.

- The O-Horizon: The O horizon is the upper layer of the topsoil which is mainly composed of organic materials such as dried leaves, grasses, dead leaves, small rocks, twigs, surface organisms, fallen trees, and other decomposed organic matter. This horizon of soil is often black brown or dark brown in colour and this is mainly because of the presence of organic content.

- The A-Horizon or Topsoil: This layer is rich in organic material and is known as the humus layer. This layer consists of both organic matter and other decomposed materials. The topsoil is soft and porous to hold enough air and water. In this layer, the seed germination takes place and new roots are produced which grows into a new plant. This layer consists of microorganisms such as earthworms, fungi, bacteria, etc.
- The E-Horizon: This layer is composed of nutrients leached from the O and A horizons. This layer is more common in forested areas and has lower clay content.
- The B-Horizon or Subsoil: It is the subsurface horizon, present just below the topsoil and above the bedrock. It is comparatively harder and compact than topsoil. It contains less humus, soluble minerals, and organic matter. It is a site of deposition of certain minerals and metal salts such as iron oxide. This layer holds enough water than the topsoil and is lighter brown due to the presence of clay soil. The soil of horizon-A and horizon-B is often mixed while ploughing the fields.
- The C-Horizon or Saprolite: This layer is devoid of any organic matter and is made up of broken bedrock. This layer is also known as saprolite. The geological material present in this zone is cemented.
- The R-Horizon: It is a compacted and cemented layer. Different types of rocks such as granite, basalt and limestone are found here. Apart from the rocks, minerals, and layers, soil profile also consists of a water content, which is referred to as soil moisture.

Soil Moisture

Water in the soil is referred to as soil moisture. Water absorption in the soil is determined by various factors. It plays a major role in soil formation. As a result of precipitation, water arrives at the surface. The particle size distribution of soil determines its porous nature and causes downward movement of water vertically which is known as infiltration. This penetration continues deep in the layers of soil until it reaches saturation.

Water, on reaching this barrier, cannot seep vertically further, hence moves sideways. Formation of puddles as a result of saturation is called surface ponding which can be long-lasting. Water that is available to plants is called Root zone moisture while surface soil moisture is the water available in the immediate upper region of soil.

Moisture content in the soil can be measured using a device known as Tensiometer. They are water-filled tubes which are sealed with a porous ceramic tip towards the bottom and a gauge at the top which is devoid of air molecules. They are penetrated into the soil till the root level. Water passes between the tip of the device and the ambient soil until it reaches an equilibrium and hence, tension is recorded on the gauge. Readings thus obtained give a measure of soil moisture in that region.

Types of Soil Moisture

The different types of water present in the soil include:

- Gravitational Water: The water that reaches the water table of the soil due to the gravitational force is referred to as gravitational water. This is not available to the plants.
- Hygroscopic Water: This water is also not available to the plants. It is a thin film of water tightly held by the soil particles.
- Chemically Combined Water: The chemical compounds present in the soil particles contain water. This is known as chemically combined water. This is also not available to the plants.
- Capillary Water: This water is available to the plants for absorption. This water exists between soil particles in small capillaries.
- Atmospheric Humidity: The hanging roots of the epiphytes absorb the moisture in the air due to the presence of hygroscopic hairs and spongy velamen tissues.

Importance of Soil Moisture Content

- Soil water carries food nutrients for the growth of plants.
- Soil moisture content determines the yield of the crop in a region.
- Crucial in maintaining soil's temperature.
- Soil moisture acts as nutrients.
- Important for soil formation.
- Moist soil is ideal for the growth of many plants that demand a huge supply of water (Ex: Rice).
- Soil moisture catalyses biological activities of microbes in the soil.
- Water is a primary need for photosynthesis in plants.

Measuring Soil Moisture

The soil moisture can be measured by various tools mentioned below:

- Tensiometers: These measure the tension of soil moisture. They are water-filled tubes, with a porous ceramic tip at the bottom. These are sealed and have a vacuum gauge at the top. They are inserted in the soil to the depth of the plant root zone. The readings obtained in the tensiometers indicate the availability of water in the soil.
- Electrical Resistance Blocks: These consist of two electrodes connected to lead wires extending to the soil surface. The electrodes are embedded in the blocks of porous material. It is used to measure soil water tension.
- Time Domain Reflectometry (TDR): TDR – Time Domain Reflectometry is used to determine the soil moisture content. Steel rods are placed in the soil and electrical signals are sent through them. The returned signals are measured to determine soil water content.

Soil Humus

Humus comprises of two parts: Humic and Fulvic acids (base soluble) and humin (base insoluble). Humification refers to the process by which humus is formed. During process, N/C ratio increases as carbon is lost to CO_2 evolved during biodegradation and nitrogen fixed by nitrogen fixing bacteria is incorporated into the humic residue. Humic substances exert a strong influential impact on the soil properties. They are known to bind metals strongly and also enhance water binding capacity of soil. They are also known to act as buffers in soil due to their strong acid-base character. They enhance the water holding capacity of soil. They also stabilize to increase the sorption of organic compounds by soil.

Types of Humus

Humus occurs in soils in many types, differentiate in regard to morphology and fractional composition. A type of humus is it a morphological form of naturals accumulation of humic substances in profile or on the surface of soil, conditioned by general direction of soil-forming process and humification of organic matter. Types of humus in terrestrial environment are:

- Mor
- Moder
- Mull

Mor

Mor is a type of humus, which occur largely in coniferous forest soils and the moorlands soils. This humus arise under conditions of low-biological activity in soil. The mineralizations of organic matter proceed slowly and create layers, which maintain a structure of vegetable material. Acidophilic fungi and low active invertebrates participate in transformations of plant residues.

Under these circumstances forms a litter of large thickness. C/N ratio of mor humus is always more than 20, or even 30-40, whereas pH is acid.

Moder

Moder is a transitional form of humus between mull and moder, characteristic for sod-podzolic soils, losses and mountain grassland soils. The organic horizons with moder humus consist of low-thicknesses litters (2-3 cm), which gradually, without bounds, pass on to humus-accumulative horizons.

Moder is a type of medium humified humus. Acidophilic fungi and arthropodan participates in transformations of plant residues. C/N ratio equal 15-25. Produced mineral-organic complexes are labile and weakly bounded with mineral portion of soil.

Mull

Mull is a type of humus characteristic for chestnut soils, phaeozems, rendzinas and others soils. This type of humus arise under grass vegetation. Mull is a well humified organic matter, which is produce in very biologically active habitat. This type of humus is characterized by neutral pH, C/N ratio nearing to 10 and ability to creation stable mineral-organic complexes.

Mull is a type of humus which occurs in soils under cultivation. According to Kononova, the types of humus are divide as follows:

- First type of humus is characteristic for podzolic soils, grey brown soils and lateric soils under forest communities. In this humus predominate humic acids, thus humic acid/fulvic acid ratio is below 1. Humic acid indicate small extent of aromatic rings condensation and they are approximate to fulvic acids. Considerable hydrophilic properties of humic acids favor to creation of chelates with polyvalent cations and ability to displacement deep into profile of soil. Considerable mobility of this humus favor process of podsolization.
- Second type of humus is characteristic for phaeozems, rendzinas, black earths and brown soils. Humic acid/fulvic acid ratio is upper than 1, Extent of aromatic rings condensation is high in humic acids, which cause their hydrophobic properties and inability to creation of chelates. Humic acids are strongly connected with mineral portion of soil in this type of humus.
- Third type of humus is characteristic for semi-desert soils. In this humus predominate fulvic acids fraction, whereas arise of humic acids is limited. Beyond this, humic acids are largely bounded with mineral portion of soil.

Soil Stabilization

Soil stabilization is a method of improving soil properties by blending and mixing other materials. Improvements include increasing the dry unit weight, bearing capabilities, volume changes, the performance of in situ sub-soils, sands, and other waste materials in order to strengthen road surfaces and other geotechnical applications.

Stabilization using Cement

One of the common methods of chemical stabilization is to mix soil with cement to form a product named as soil–cement. Soil–cement can be defined as a mixture of soil and measured amounts of Portland cement and water and compacted to the desired density. Soil–cement has been used as a base material as an adoption of improved measure in many projects, such as slope protection of dams and embankments, pavement of highways, building pads, terminals for rail and truck, composting facilities, cheap base for streets, parking lots, channels and reservoir linings, mass soil–cement placement for dikes, foundation stabilization etc. The soil–cement technique has been practiced almost for 100 years. It serves to amend the mechanical and the engineering properties of the soil. The new performance depends on the ability of the additives to react with the mixing soil. There are four main properties of soil; strength, permeability, volume stability, and durability that can be enhanced with additives. The choice of a specific additive depends on the type of soil, service that is required to serve and the surrounding environment. When water is mixed with cement, hydration occurs, meaning cementing compounds of calcium–silicate–hydrate (C–S–H) and calcium–aluminate–hydrate (C–A–H) are formed and excess calcium hydroxide (CaOH) is released, approximately 31% by weight. Formation of C–S–H and C–A–H occurs when crystals begin forming a few hours after the water and cement are mixed; crystals will continue to form as long as unreacted cement particles and free water remain within the mixture. Five standard types of Portland cement (Types I through V) are available as specified by ASTM C150.

The process of cementation and the results of soil–cement and lime stabilization are similar, they used in quantities too small to provide high-strength cementing action. They reduce the plasticity of clay soils. Calcium chloride or sodium chloride are added to the soil to retain moisture and also control dust, to hold fine material for better compaction, and to reduce frost heave by lowering the freezing point of water in the soil. Kezdi reports that cement treatment slightly increases the maximum dry density of sand and highly plastic clays but it decreases the maximum dry density of silt. In contrast studies by Deng and Tabatabai shows that cement increases the optimum water content but decreases the maximum dry density of sandy soils. Cement increases plastic limit and reduces liquid limit, which mainly reduces plasticity index. The other significant effects of soil–cement stabilization is reduction in shrink-age and swell potential, increase in strength, elastic modulus, and resistance against the effect of moisture, freeze, and thaw. Cement treated soils show a brittle behavior compare of non-treated soils. Cement can be applied to stabilize any type of soil, except soils with organic content greater than 2% or having pH lower than 5.3. The use of cement in granular soils has proven to be economical and effective because smaller amounts of cement are required. In addition, soils that have a PI value higher than 30 are difficult to mix with cement. To avoid this issue, lime can be added prior to mixing in cement; this initial step will keep soils more workable.

Khemissa and Mahamedi found that swell pressure decrease as the stabilizer content increased in cement treated samples. Cementitious links develop between the calcium silicate and calcium aluminate found in Portland cement with the soil particles. Unlike lime, hydration in cement occurs at a faster pace which allows for an immediate strength gain. Therefore, there is no need of a mellowing period when stabilizing with cement; compaction of soil–cement samples is typically conducted within 2 h of initial mixing. The strength gain achieved during compaction may be below the ultimate strength of a soil–cement sample. However, the cement stabilized soil will continue to gain strength over the course of several days.

There are many factors contributing to the length of curing time required for strength gain in soil–cement samples. These include ambient air temperature, relative humidity, type of cement used, and concentration of cement used. Guthrie and Reese found that the relative strength is sensitive to the factors, while the relative compaction is not. Faster wind speed, higher air temperature, lower relative humidity and longer delay in compaction commonly result in a poor strength.

Stabilization using Lime

Limestone is broken down at elevated temperatures to form lime. As a result, three forms of lime are produced: quicklime (calcium oxide–CaO), hydrated lime (calcium hydroxide–$Ca[OH]_2$), and hydrated lime slurry; all of which can be used to treat soils. Quicklime is manufactured by chemically transforming calcium carbonate (limestone– $CaCO_3$) into calcium oxide. Furthermore, hydrated lime is created when quicklime chemically reacts with water. When hydrated lime is mixed with clay particles, it permanently forms strong cementitious bonds. Lime has been known to reduce the swelling potential, liquid limit, plasticity index and maximum dry density of the soil, and increases its optimum water content, shrinkage limit and strength. It improves the workability and compact ability of subgrade soils.

If quicklime is used, it immediately hydrates (i.e., chemically combines with water) and releases heat. Soils are dried, because water present in the soil participates in this reaction, and because the heat generated can evaporate additional moisture. The hydrated lime produced by these initial reactions will subsequently react with clay particles. These subsequent reactions will slowly produce additional drying because they reduce the soil's moisture holding capacity. If hydrated lime or hydrated lime slurry is used instead of quicklime, drying occurs only through the chemical changes in the soil that reduce its capacity to hold water and increase its stability. After initial mixing, the calcium ions (Ca^{2+}) from hydrated lime migrate to the surface of the clay particles and displace water and other ions. The soil becomes friable and granular, making it easier to work and com- pact. At this stage the Plasticity Index of the soil decreases dramatically, as does its tendency to swell and shrink. The process, which is called "flocculation and agglomeration," generally occurs in a matter of hours.

When adequate quantities of lime and water are added, the pH of the soil quickly increases to above 10.5, which enables the clay particles to break down. Silica and alumina are released and react with calcium from the lime to form calcium-silicate-hydrates (CSH) and calcium-aluminate-hydrates (CAH). CSA and CAH are cementitious products similar to those formed in Portland cement. They form the matrix that contributes to the strength of lime-stabilized soil layers. As this matrix forms, the soil is transformed from a sandy, granular material to a hard, relatively impermeable layer with significant load bearing capacity. The matrix formed is permanent, durable, and significantly impermeable, producing a structural layer that is both strong and flexible. However, alteration of particle structure occurs slowly, depending upon the type of clay present, a mellowing period from 1 to 4 days is allowed to obtain a homogeneous, friable mixture. These reactions maybe summarized as:

$$Ca^{2+} + OH^- + \text{Soluble Clay Silica} \rightarrow \text{Calcium Silicate Hydrate (CSH)}$$

$$Ca^{2+} + OH^- + \text{Soluble Clay Alumina} \rightarrow \text{Calcium Alumina Hydrate (CAH)}$$

Lime stabilization enhances engineering properties of soils, such as improved strength, higher resistance to fracture, fatigue, and permanent deformation, enhanced resilient properties, reduction in swelling; and resistance to the harmful effects of moisture. The most considerable improvements in these properties are observed in moderately to highly plastic clays.

Al-Kiki et al. acknowledged that over the time, the properties of treated soil affect the strength gain. Soil pH, organic content, the quantity of exchangeable sodium, clay mineralogy, natural drainage, weathering conditions, extractable iron, carbonates and silica-alumina ratio are some of the properties which influence the gain in strength. The stabilization of acidic soil using lime, resulted in lower compressive strength than that of alkaline soil. Broderick and Daniel reported that the lime and cement stabilized soils are less vulnerable to attack by organic chemicals in comparison to untreated soils. Haraguchi et al. investigated the variation of the engineering properties of freshly cement-stabilized decomposed granite soil cured in water and in 0.2 N acid solution, and indicated that the CBR obtained from the specimens cured in the 0.2 N acid solution was lower than that cured in water.

The strong alkaline conditions were able to release silica and alumina from the clay mineral and eventually react with lime to form new cementation products. The success of the lime treatment process is highly dependent on the available lime content, curing time, soil type, soil pH and clay minerals. Limited research has been conducted to determine whether pH variations will affect properties of lime-stabilized soils. Additional studies are therefore necessary to explain the erosion mechanism of lime-stabilized soils due to pH variations. However, experience shows that lime will react with many medium-, moderately fine- and fine- grained soils to produce decreased plasticity, increased workability, reduced swell, and increased strength. Soils classified according to the USCS as CH, CL, MH, ML, OH, OL, SC, SM, GC, GM, SW-SC, SP-SC, SM-SC, GWGC, GP-GC, ML-CL, and GM-GC should be considered as potentially capable of being stabilized with lime. Lime should be considered with all soils having a PI greater than 10 and more than 25 percent of the soil passing the No. 200 sieve.

Stabilization using Fly Ash

Fly ash is one of four coal combustion products (CCPs) that are produced as a by-product of burning coal, two major groups, Class C and Class F fly ash are produced. Burning lignite and subbituminous coal produces Class C fly ash. However, burning anthracite, other known as bituminous coal, produces Class F fly ash. Although there can be multiple variations of the chemical additive, fly ash particles generally consist of hollow spheres of silicon, aluminum, and iron oxides and unoxidized carbon all of which make both classes of fly ash pozzolans-siliceous or siliceous and aluminous materials.

It is typically viewed as non-plastic fine silt (ML) when using the Unified Soil Classification System (USCS). Class F fly ash is not used as often because it requires an activator, either lime or cement, to form pozzolanic stabilized mixtures (PSMs) since it is not a self-cementing material. Zulkifley et al. assessed the effect of off- specification fly ash on the engineering properties of tropical soils from Hawaii. They observed that the fly ash reduced the liquid limit and plasticity index, and enhanced the California bearing ratio (CBR) and unconfined compressive strength. Radhakrishnan et al. performed unconfined compressive strength (UCS) and one-dimensional free swell tests on soil–lime mixtures altered with Class F fly ash. Observations showed that lime and fly ash are a good combination for stabilizing silty and sandy soils. It drastically increased the stiffness of the final product. Rupnow et al. used Class C fly ash with lime to develop guidelines for estimating the coefficient of structural layer for the base layer of flexible pavement. The required base layer thickness decreased with the addition of both additives. Fly ash, when mixed with lime, can be used effectively to stabilize most coarse- and medium-grained soils; however, the PI should not be greater than 25. Soils classified by the USCS as SW, SP, SP-SC, SW-SC, SW-SM, GW, GP, GP-GC, GW-GC, GP-GM, GW-GM, GC-GM, and SC-SM can be stabilized with fly ash.

Soil Stabilization using Fibers

The use of hair-sized polypropylene fibers in soil stabilization requests has been common in soil stabilization projects for its low cost compared with other stabilization agents. These materials have a high resistance towards chemical and biological degradation and do not cause leaching in the soil. Sharma et al. studied the improvement in the properties of swelling clay using hay fibers. They found that there was no considerable or reasonable change in the Atterberg limits due to introduction of hay fiber. The maximum dry density (MDD) decreased with the addition of hay. The optimum moisture content (OMC) decreased with increasing hay content up-to 1.0% then began to decrease. The shrinkage limit decreased with increasing hay content up-to 1.0% then increased. The unconfined compression strength decreased with increasing of hay content to soil ratio. The direct shear strength increased notably with hay addition. The tensile strength of the air dried mixture increased with hay addition and the swelling decreased with hay addition.

Mirzababaei et al. studied unconfined compressive strength (UCS) of reinforced clays with the waste of carpet fibers. They found that carpet waste fibers mixed with clay soils, prepared at the same dry unit weight, can considerably increase the unconfined compression strength (UCS), decrease the post peak strength loss, and alter the failure behavior from brittle to ductile. The results also indicated that the relative benefit of fibers to enhance the UCS of the clay soils is largely dependent on initial dry unit weight and water content of the soil. Cristelo et al. studied the performance of fiber rein- forced clayey soil.

They found that the excessively compressible clayey soil reinforced with randomly mixed fibers, resulted in higher ultimate bearing capacity, and lesser settlement at the ultimate load. Yilmaz in studying the compaction characteristics of reinforced soil, found that the addition of fiber, affected compaction behavior of samples. The increment in fiber content and length of fiber caused increment in optimum moisture content (OMC) and slight decrement of maximum dry density (MDD). Anagnostopoulos et al. found that with the inclusion of fibers into the soil, the unconfined compressive strength and the corresponding strain at failure increase up to an optimum fiber content and length, and decrease thereafter. In addition; inclusion of fibers in soil increases the strain at failure, and therefore makes the reinforced soil matrix more duc- tile. The settlement under a particular load in unreinforced soil is much more than the reinforced soil. Minimum settlement being observed for the soil reinforced with poly- propylene fibers.

Anggraini et al. studied the stiffness and strength response of coir fiber-reinforced tropical soil. They found that the stress–strain behavior of soil improved by incorporating coir fibers into the soil. The maximum increase in strength was observed when the fiber length is between 15 and 25 mm. The stiffness of soil increases and also the immediate settlement of soil reduced considerably due to the inclusion of fibers in the soil. Shukla et al. studied analytical model of granular soils reinforced with fibers, under high confining pressures. They found that the apparent cohesion and shear strength increase is proportional to the fiber content and aspect ratio.

The increase in shear strength of the granular soil stabilized with fibers was mostly from the apparent cohesion, and the contribution from the increase in normal confining stress was quite limited. Estabragh et al. studied the mechanical behavior of a fiber-clay composite with natural fiber. They found that the inclusion of the fiber reinforcement within the soil causes a decrease in pre-consolidation stress and increases compressibility and swelling indices. In addition, the strength and angle of internal friction increase considerably for the total and effective stresses.

Mechanisms of Stabilization

Mechanisms of stabilization that utilize cement, lime, or fly ash were summarized by Little as follows:

- Cation exchange: Sodium, magnesium, and other cations are replaced by the calcium cations from the available calcium hydroxide.
- Flocculation and agglomeration: Flocculation of the clay particles increases the effective grain size and reduces plasticity, thus increasing the strength of the matrix.
- Pozzolanic reaction: The high pH environment created by the available calcium hydroxide solubilizes silicates and aluminates at the clay surface, which in turn react with calcium ions to form cementitious products that are composed primarily of calcium silicate hydrates or calcium aluminate hydrates, or both.
- Carbonate cementation: Calcium oxide reacts with carbon dioxide from the atmosphere to form calcium carbonate precipitates, which cement the soil particles.

Cementitious hydration reaction, where calcium silicates and/or calcium aluminates, which are chemically combined in the production of Portland cement clinker or in the coal burning (fly ash) process, hydrate rapidly (within a few hours) for calcium silicate and/or calcium aluminate hydrates. The chemical reactions during the hydration procedure include, but are not limited to, some or all of the reactions listed below:

$$2C_2S + 6H_2O \rightarrow C_3S_2H_3 + 3Ca(OH)_2$$

$$2C_2S + 4H_2O \rightarrow C_3S_2H_3 + Ca(OH)_2$$

$$C_3A + 3(CaSO_4 2H_2O) \rightarrow 26H_2O \rightarrow C_3A \cdot 3CaSO_4 \cdot 32H_2O$$

$$2C_3A + C_3A3\ CaSO_4 32H_2O \rightarrow 3[C_3ACaSO_4 12H_2O]$$

$$C_3S + Ca(OH)_2 + 12H_2O \rightarrow C_3A\ Ca(OH)_2\ 12H_2O$$

$$C_4AF + 3(CaSO_4 2H_2O) + 27H_2O \rightarrow C_3(AF)3CaSO_4\ 32H_2O + Ca(OH)_2$$

$$2C_4AF + C_3(AF)3CaSO_4\ 32H_2O + 6H_2O \rightarrow 3[C_3(AF)CaSO_4 12H_2O] + 2Ca(OH)_2$$

$$C_4AF + 10H_2O + 2Ca(OH)_2 \rightarrow C_3AH_6 - C_3FH_6 \text{ (Solid Solution)}$$

Two popular theories on the mechanisms of stabilization, namely the crystalline theory by Le Chatelier, and the gel theory proposed by Taylor, have been integrated into a combined gel/crystalline theory that describe the different stages of curing. Li provided a detailed description of the hydration process of cement concrete: If a proper amount of moisture is available to ensure complete hydration of the cement, then each cement particle is covered with water and thus forms a gel-like film. These coated particles of cement coat the aggregate or soil particles as well. At this stage, the cement has not begun to set. The reaction between the water and the cement forms small single crystals. With the development of hydration, the single crystals begin to grow into one another and form a crystalline network. The adjoining crystals are attracted to one another by van der Waals forces instead of a chemical bond. Figure illustrates this process:

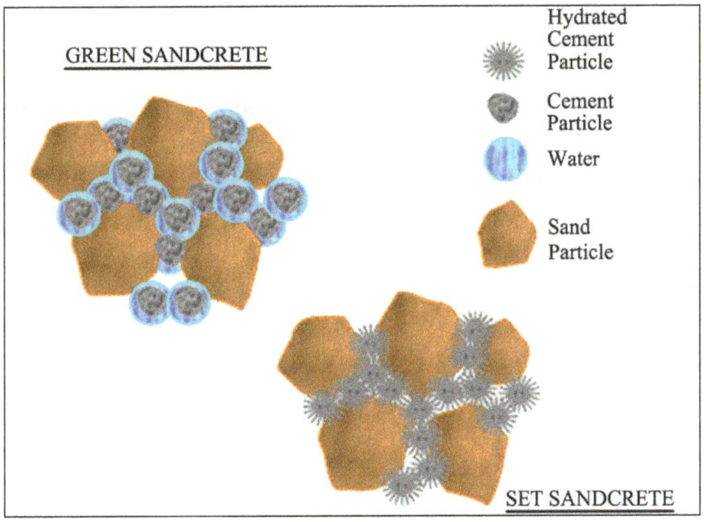

Bonding process between cement particles and sand particles.

Advantages and Disadvantages of Calcium–Based Additives

Soil stabilization using calcium-based additives involve advantages and disadvantages. Some of the advantages and disadvantages of calcium-based additives are discussed here.

Advantages

The following list presents some of the engineering advantages of calcium-based soil stabilization:

- Functions as a working platform (expedites construction).
- Reduces the need for dusting.
- Waterproofs the soil.
- Upgrades marginal aggregates or soils.
- Improves strength.
- Improves durability.
- Controls volume changes of soils.
- Improves soil workability.
- Dries wet soils.
- Reduces pavement thickness requirements.
- Conserves aggregate.
- Reduces construction and haul costs.
- Conserves energy.
- Provides a temporary or permanent wearing surface.

In spite of the significant advantages, many serious disadvantages are inherent from calcium-based stabilizers. These disadvantages are as follows:

Negative Environmental Impact

Global warming is a serious threat which our planet is facing. Carbon dioxide (CO_2) is one of the main factors for this warming. Cement manufacturing is a process which emits CO_2 in large amount. Cement industry alone produces about 10% of total CO_2 emission. Cement manufacturing emits CO_2 through decarbonisation of limestone, burning fossil fuels, electricity, and transportation.

Sulfate Attack and Carbonation

Two deleterious chemical reactions occur in the soils treated with calcium-based materials. One of the sulfate salts existing in the soil and two lime carbonation. In the soils containing sulfates, any calcium-based additives results in heaving and disintegration leading to a loss in strength. Soil composition, groundwater and mixed water can be the source of sulfates. Chemical reaction between calcium and aluminium present in the soil mineralogy in the company of soluble sulfate and water produce ettringite and/or thaumasite. Ettringite is a hydrous calcium aluminium sulfate mineral with formula: $Ca_6Al_2(SO_4)3(OH)_{12} \cdot 26H_2O$. It is a colorless to yellow mineral crystallizing in the trigonal system. The prismatic crystals are typically colorless, turning white on partial dehydration. Calcium sulfate sources, such as gypsum, are intentionally added to Portland cement to regulate early hydration reactions to prevent flash setting, improve strength development, and reduce drying shrinkage. Sulfate and aluminate are also present in supplementary cementitious materials and admixtures. Gypsum and other sulfate compounds react with calcium aluminate in the cement to form ettringite within the first few hours after mixing with water. Most of the sulfate in the cement is normally consumed to form ettringite at early ages. The formation of ettringite in the fresh, plastic concrete is the mechanism that controls stiffening. At this stage ettringite is uniformly and discretely dispersed throughout the cement paste at a sub-microscopic level (less than a micrometer in cross-section). Taumasite is a silicate mineral with chemical formula $Ca_3Si(OH)6(CO_3) \cdot (SO_4) \cdot 12H_2O$. It occurs as colourless to white prismatic hexagonal crystals, typically as acicular radiating groups. It also occurs as fibrous masses. Its Mohs hardness is 3.5 and it has a specific gravity of 1.88–1.90.

Effect of Organic Materials

Chan and Heenan suggested that the high microbial biomass present in the organic soil activates rates of decomposition in organic soils treated with lime resulting into reduction in pH value. In addition, clay minerals are in lesser quantity in the organic soil. Therefore, the organic materials slow down the pozzolanic reaction necessary to attain soil strength. The organic soil is characterized by high water with holding capacity that may lead to a reduction in available water for the hydration process. Furthermore, organic materials coat the additive particles thus holding up the hydration process.

According to Morse et al., there is a great tendency for the reaction between calcium ions and black humic acid generated from lime to produce insoluble calcium humic acid. Hossain et al. indicated that the decomposition of organic materials hinders the polymerization of silicate. Thus, the cementitious formation is blocked by organic materials. Though, it is key to know that not all of the organic materials block cementitious reaction. Some have no effects such as chloronaphthalene and others retard hydration reaction but do not affect soil strength gain such as ethylene glycol, benzoic acid, cellulose.

Cement-treated soil to face the similar situation as mentioned for soils treated with lime. Sasanian and Newson and Zhang et al. observed that the soil treated with cement is susceptible to organic materials hindering the chemical reaction which produces cementitious materials. Huat et al. reported that soil containing organic acids and having pH less than 9 inhibit the cementing reaction. For the purpose to reduce the negative effect of organic materials, bentonite is added to the organic soil. This treatment provides two benefits, one, bentonite has high water retention ability, which is favorable for lime hydration. Two, bentonite is the source of silica for pozzolanic reaction and can become a filler. Kaolinite and zeolite may also be used for lime treated soil having humic acid. This gives enough amount of silica that is required for pozzolanic process.

Problems in Sulfate Soils Stabilized With Lime/Cement

Though lime stabilization improves the volume and strength characteristics of the expansive soils, there are some limitations to lime stabilization. These limitations include the presence of organic carbon and soluble sulfates. It has been reported that the presence of organic carbon in excess of one percent can interfere with the pozzolanic reactions, leading to low strength gains. Compared to organic carbon, the presence of sulfates is of higher concern because lime treatment in these types of soils leads to excessive heaving and pavement failures. It has been reported that when soils contain sulfate minerals such as gypsum ($CaSO_4{\cdot}2H_2O$) and sodium sulfate (Na_2SO_4) in their natural formation and are treated with calcium-based stabilizers, adverse reactions occur, causing severe heave and pavement distress. These adverse reactions are due to formation of expansive minerals, ettringite $Ca_6 \cdot [Al(OH)_6]_2 \cdot (SO_4)_3 \cdot 26H_2O$, and thaumasite $Ca_6 \cdot [Si(OH)_6]_2 \cdot (SiO_4) \cdot (CO_3)_2 \cdot 24H_2O$. This phenomenon was first reported by Sherwood. There are several sources of sulfates in soils, produced from primary or secondary origin. Primary sources can be defined as the direct sources of sulfates in their natural form, as sulfate-bearing minerals such as gypsum while, the secondary sources are those that are not a direct source of sulfate but give out sulfates as a by-product of oxidation or other forms of chemical interactions. The following section highlights these sources.

Recent studies have revealed that the soil containing sulfates, when treated with lime or cement undergo excessive swelling and pavement failures. These failures are due to the formation of ettringite. Reactions between lime, alumina, and sulfates in the soils produce an expansive crystal- line mineral called ettringite $Ca_6 \cdot [Al(OH)_6]_2 \cdot (SO_4)_3 \cdot 26H_2O$, and thaumasite $Ca_6 \cdot [Si(OH)_6]_2 \cdot (SiO_4).(CO_3)_2 \cdot 24H_2O$. Generally, it is recommended that for the sulfate enriched soils, lime and cement treatments should be chosen cautiously.

Soil–Lime–Sulphate Reactions

Lime–soil reactions are grouped into two, short term and long term reactions. Short term reactions are flocculation, pH and cation exchange reactions, lime migration carbonation, and these affect the physical properties of the soil such as consistency limits and grain size distribution. The long-term pozzolanic reactions consist of formation of various products, resulting in the growth new materials and affect the strength and compressibility of clays. Takemoto and Uchikawa explained the reaction mechanism of C_3S pozzolanic compounds, and ettringite formation through pozzolanic reactions, in a schematic way as shown in Figure.

Takemoto and Uchikawa proposed schematic explanations of the C_3S-pozzolan reaction and C_3A-pozzolan reaction, which are shown in Figure, respectively. In C_3S-pozzolan system, calcium ions dissolved from C_3S route around freely in liquid and are adsorbed on the surfaces of pozzolan particles. C–S–H made by the hydration of C_3S precipitate as the hydrates of high Ca/Si ratio on the surface of C_3S grains and as the porous hydrates of low Ca/Si ratio on the surfaces of pozzolan particles. Attack of the pozzolan surface in water brings about gradual dis- solution of Na^+ and K^+, resulting in Si and Al rich amorphous layer on the surfaces. Dissolved Na^+ and K^+ raise the OH^- concentration and speed up the dissolution of SiO_4^{-4} and layer. Due to the osmotic pressure, the layer swells slowly and the void between layer and pozzolan particle is formed. When the pressure in the void breaks the film, SiO_4^{-4} and AlO_2^- diffuse into the Ca^{2+} rich solution. In addition, C–S–H and Ca–Al hydrate precipitate on the surface of outer hydrates of C_3S particles and to a slight degree on the broken film. Vacant space remains inside the film as the hydrates do not precipitate there because of high concentration of alkalies. For pozzolans with low alkalies, damage of amorphous Si, Al rich film enables Ca^{2+} to travel into the inside of the film and precipitate calcium silicate and calcium aluminate hydrates on the surface of pozzolan grain. Therefore, no space is witnessed between pozzolan grains and hydrates.

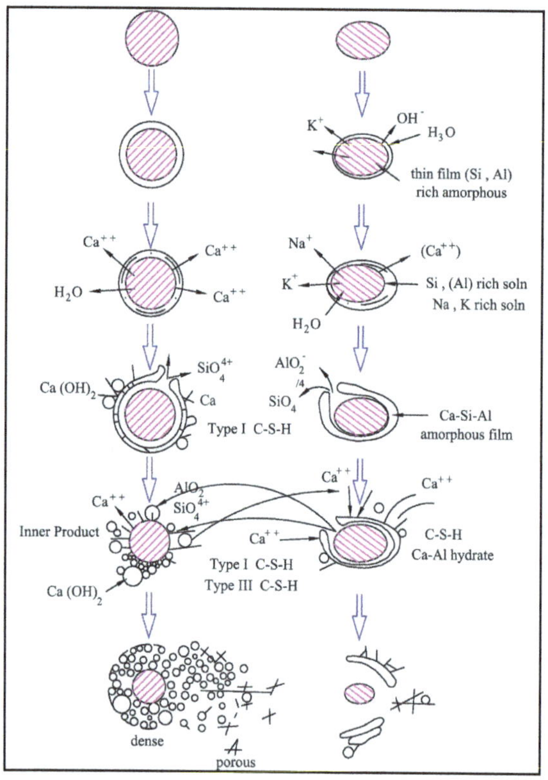

Reaction mechanism involved in the development of C3S-Pozzolana grains.

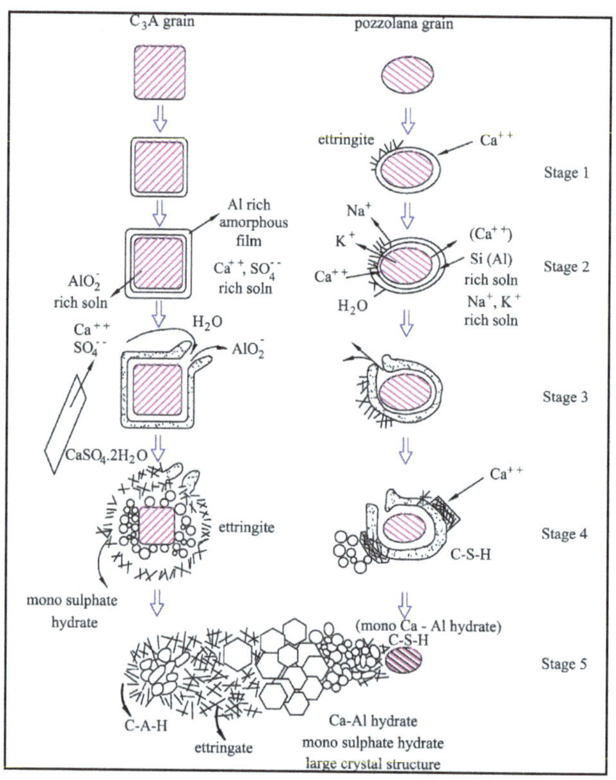

Schematic development of ettringite formation.

The hydration of the C_3A-pozzolan system was first observed by Uchikawa and Uchida. Figure illustrates the schematic development of the C_3A-pozzoloan system in the presence of calcium hydroxide and gypsum. Uchikawa and Uchida, pointed out that the presence of the pozzolan speeds up the hydration of C_3A by adsorbing Ca^{2+} from the liquid phase and by providing precipitation sites for ettringite and other hydrates. The C_3A-pozzolan reaction system is comparable to Aluminate hydrate, and calcium silicate hydrate are shaped on the surface film outside the pozzolan particles of on the surface hydrate layer of the C_3A particles, depending on the concentration of Ca^{2+} and SiO_4^{-2} in solution.

Soil Fauna Diversity

Soil represents one of the most important reservoirs of biodiversity. It reflects ecosystem metabolism since all the bio-geo-chemical processes of the different ecosystem components are combined within it; therefore soil quality fluctuations are considered to be a suitable criterion for evaluating the long-term sustainability of ecosystems. Within the complex structure of soil, biotic and abiotic components interact closely in controlling the organic degradation of matter and the nutrient recycling processes. Soil fauna is an important reservoir of biodiversity and plays an essential role in several soil ecosystem functions; furthermore, it is often used to provide soil quality indicators. Although biodiversity was one of the focal points of the Rio conference, in the 1990s virtually no attention was paid to activities for the conservation of soil communities.

Human activities frequently cause a degradation of soil environmental conditions which leads to a reduction in the abundance and to a simplification of animal and plant communities, where species able to bear stress predominate and rare taxa decrease in abundance or disappear. The result of this biodiversity reduction is an artificial ecosystem that requires constant human intervention and extra running costs, whereas natural ecosystems are regulated by plant and animal communities through flows of energy and nutrients, a form of control progressively being lost with agricultural intensification. For these reasons the identification of agricultural systems which allow the combination of production targets and environmentally friendly management practices, protecting both soil and biodiversity, is essential in order to prevent the decline of soil fauna communities in agricultural landscapes.

Biodiversity of Soil Fauna

Soil biota are thought to harbour a large part of the world's biodiversity and to govern processes that are regarded as globally important components in the recycling of organic matter, energy and nutrients. Moreover, they are also key players in several supporting and regulating ecosystem services. Furthermore, they are key components of soil food webs. Rough estimates of soil biodiversity indicate several thousand invertebrate species per site, as well as the relatively unknown levels of microbial and protozoan diversity. Soil ecosystems generally contain a large variety of animals, such as nematodes, microarthropods such as mites and Collembola, Symphyla, Chilopoda, Pauropoda, enchytraeids and earthworms. In addition, a large number of meso- and macrofauna species (mainly arthropods such as beetles, spiders, diplopods, chilopods and pseudoscorpion, as well as snails) live in the uppermost soil layers, the soil surface and the litter layer.

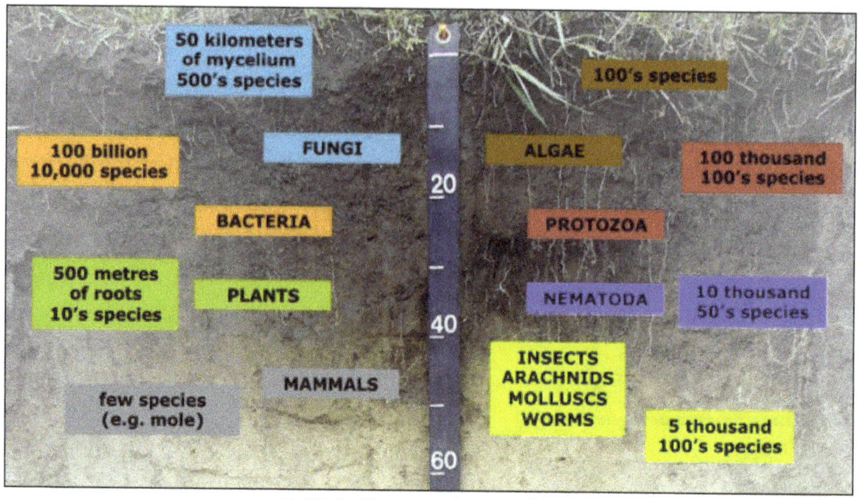

Soil Biodiversity in Numbers.

Earthworm belonging to megafauna.

Pseudoscorpion, a typical organism that live in the wood litter.

In general, soil invertebrates are classified according to their size in microfauna, mesofauna, macrofauna and megafauna.

- Microfauna: organisms whose body size is between 20 µm and 200 µm. Just one group, protozoa, is found wholly within this category; among the others, small mites, nematodes, rotifers, tardigrades and copepod crustaceans all fall within the upper limit.
- Mesofauna: organisms whose body size is between 200 µm and 2 mm. Microarthropods such as mites and springtails, are the main representatives of this group, which also includes nematodes, rotifers, tardigrades, small araneidae, pseudoscorpions, opiliones, enchytraeids, insect larvae, small isopods and myriapods.
- Macrofauna: organisms whose size is between 2 mm and 20 mm. This category includes certain earthworms, gastropods, isopods, myriapods, some araneidae and the majority of insects.
- Megafauna: organisms whose size exceeds 20 mm. The members of this category include large size invertebrates (earthworms, snails, myriapods) and vertebrates (insectivores, small rodents, reptiles and amphibians).

Despite several decades of soil biological studies it is still very difficult to provide average abundance and biomass values for soil invertebrates. This is partly caused by their high variability in both time and space, as well as by differences in the sampling methods used. In addition, most work has been performed in the forest soils of temperate regions, while other eco-regions such as the tropics, or other land uses such as agriculture, have been seriously neglected.

Soil fauna are highly variable and the majorities are also highly adaptable with regard to their feeding strategies, ranging from herbivores to omnivores and including carnivores. Depending on the available food sources many soil fauna are able to change their feeding strategies to a greater or lesser extent with many carnivorous species able to feed on dead organic matter in times of low food availability. The interactions between soil fauna are numerous, complex and varied. As well as the predator/prey relationships and in some instances parasitism, commensalism also occurs.

The degree of interaction between soil organisms and the soil itself can be highly variable among taxa and dependent on the part of the life cycle that is spent in the soil. In particular, in this regard, combined with the morphological adaptations and the ecological functions of organisms, it is possible to classify soil fauna into four main groups: temporarily inactive geophiles, temporarily active geophiles, periodical geophiles, and geobionts. It should be noted that these groupings do not have any taxonomical significance but rather are useful when studying the life strategies of soil invertebrates. Temporarily inactive geophiles are organisms that live in the soil for only certain phases of their life, such as to overwinter or to undergo metamorphosis, when protection from climatic instability is more necessary. Due to their relative inactivity, the organisms belonging to this group have a weak influence on the ecological function of soil, although they can be important as prey for other organisms.

Temporarily active geophiles live in the soil in a stable manner for a large part of their life (i.e. for one or more development stages, and emerge from the soil as adults). Most of these organisms are insects, such as Neuroptera, Diptera, Coleoptera and Lepidoptera. Organisms having a "pupae" stage in their life cycle play a minor role in the soil during these phases, while the "larvae" stage is much more important for the ecology of soil, especially when the population density is high. Most larvae can act as both detritivores and predators. Periodical geophiles spend a part of their life cycle in the soil, generally as larvae, but throughout their lives they occasionally go back to the soil to perform various activities, such as hunting, laying of eggs or to escape dangers. Several Coleoptera groups (e.g. carabides, scarabeids, cicindelids) spend their larval stage in the litter or in the upper layers of mineral soil, and when adults, use soil as a food source, a refuge and for other purposes. Geobionts are organisms that are very well adapted to life in soil and cannot leave this environment, even temporarily, having characteristics that prevent survival outside of the soil environment due to their lacking protection from desiccation and temperature fluctuations, as well as the sensory organs necessary to survive above ground by finding food and avoiding predators. Several species of myriapods, isopods, Acari, molluscs and the majority of Collembola, Diplura and Protura, belong to this group. These different types of relationships between soil organisms and the soil environment determine a differentiated level of vulnerability among various groups, as a consequence of any possible impact on soil environment. For instance, if soil contamination occurs, any impact will be highest on geobyonts (because they cannot leave the soil and must spend all their life there) and lowest on temporarily inactive geophiles.

The main four grouping that can be individualized between soil invertebrates, depending on their life strategies and how closely they are linked with soil.

There are many extremely old groups of microarthropods in soils, such as collembola and mites, dating from the Devonian period (more than 350 million years ago). In relation to the origin of soil microarthropods, it is possible to form two hypothetical groups: the first group originated in epigeous (above the soil) habitats and only subsequently adapted to soil (e.g. Coleoptera, Chilopoda, Diplopoda and Diptera.

Centipede (Chilopoda).

Millipede (Diplopoda).

Polyxena, a curious group of Diplopoda.

The second group possibly originated directly in the soil. This group contains organisms such as Protura, Diplura, Symphyla, Pauropoda, and Palpigrada which do not have forms in epigeous, or aquatic habitats (some exceptions are found in caves, where the environmental conditions are very similar to that of soils).

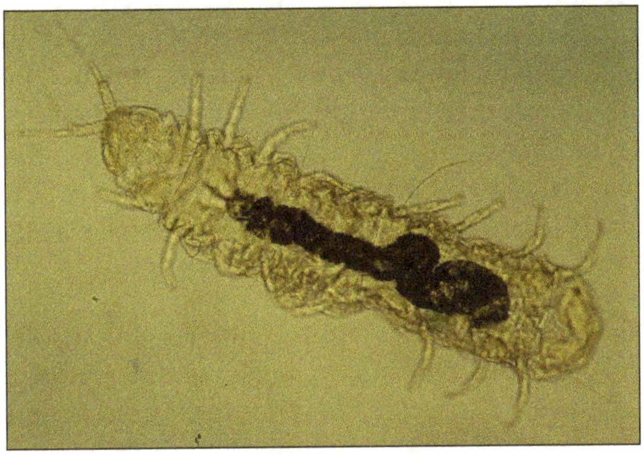

Pauropoda, a microarthropod well adapted to soil.

Over the very long period of adjustment to life below ground, the bodies of euedaphic microarthropods became adapted with characteristics that allow them to survive within the soil habitat. During this process of adaptation, impressive levels of convergence have occurred with many of the adaptation characteristics being morphological, and easily explainable and understood. For example, the reduction of the visual apparatus, loss of pigmentation or cryptal coloration (camouflage), reduction of appendages and the acquiring of special structures essential for life below ground. Some of these characteristics, such as the reduction of body length (miniaturisation), loss of the appendages (legs, antenna, etc.) and the loss of eye functionality, which in some cases leads to the complete disappearance of eyes, are direct consequences of degenerative processes in structures which are very important in above ground habitats but useless in the soil. Conversely, soil microarthropods have developed characteristics that permit them to live in the particular conditions present in the soil, such as chemico- and hydroreceptors, often distributed not only in the oral region, as they are for most above ground organisms, but also on other structures of the body. The confinement of these groups to soils, i.e. the groups' incapacity to leave them, is due to the relative stability of these habitats.

In fact, diverse factors such as water, temperature and organic matter vary only slightly over the short- and medium-term, as compared to large variations in above ground environments. In addition, there is obviously no light in soil at depths greater than a few millimetres. As a result of all of these factors combined, euedaphic microarthropods are sensitive and unable to survive abrupt variations in environmental factors. They are particularly sensitive to soil degradation and to the disturbances caused, for example, by agricultural cultivation and trampling. Collembola (springtails) represent one of the most important groups of soil microarthropods, both because of the number of species, and the number of individuals, generally present in soils. They have some characteristics that make this group very interesting and useful for studying soil evolution convergence phenomena. Furthermore, they are very useful as indicators of soil quality as their biodiversity and density are influenced by numerous soil factors (in particular organic matter and water content but also other factors such as contamination).

The Role of Soil Fauna in Soil Ecosystem Processes

Some researchers have defined edaphic fauna as a "super organism" that assumes a crucial significance due to the chemico-physical and biological processes that are rooted in the soil. Soil biota play an essential role in soil functions as they are involved in processes such as the decomposition of organic matter, the formation of humus and the nutrient cycling of many elements (nitrogen, sulphur, carbon). Moreover, edaphic fauna affect the porosity and aeration of, as well as the infiltration and distribution of organic matter within soil horizons. The ecosystem services provided by soil fauna are one of the most powerful arguments for the conservation of edaphic biodiversity. Decomposition of organic matter by soil organisms is crucial for the functioning of an ecosystem because of its substantial role in providing ecosystem services for plant growth and primary productivity.

Due to the absence of light, which makes photosynthesis unfeasible, among the organisms populating the soil we find very few real phytophages, unless we extend the definition of soil animals to surface organisms, or if we consider that pests also include fungivorous microarthropods. The activity of animals, among them typically protozoa, nematodes, rotifers, certain springtails and mites, which feed on microflora, consisting of bacteria, actinomycetes and fungi (both hyphae and spores) is of crucial importance both for regulating the density and for diffusing these micro-organisms. For example, through their faeces, springtails, which feed on fungi, can spread fungal spores that are still viable to areas as far as a few metres away from their point of origin.

The detritus food chain takes on an essential role within the soil, as it becomes the basis of the hypogean food web; in fact, many organisms such as isopods, certain myriapods, earthworms, springtails, many species of mites, and the larvae and adults of many insects feed on the vegetable and animal detritus that is deposited on the soil. For example in the soil of a temperate forest, in which the contribution of litter each year can amount to 400 g/m^2, about 250 g/m^2 are ingested by earthworms and enchytraeids, 30-40 g/m^2 by mites and 50-60 g/m^2 by springtails. Soil fauna performs a mainly mechanical action, whereas chemical degradation is essentially performed by fungi and bacteria, both free and intestinal symbionts of other organisms; furthermore during digestion, organic substances are enriched by enzymes that are dispersed in the soil along with the faeces, contributing to humification.

Earthworms are among the most important organisms in many of the soils of the world. According to an ecological and functional classification it is possible to identify three groups of earthworms that differ in size, burrowing capability, type of food and habitat: epigeic, endogeic and anecic. The epigeic, whose sizes ranges from 0.5 to 5 cm, have a red coloured body, are poorly adapted to burrowing and have a good tolerance to low pH values. They inhabit the superficial organic layers where they feed on litter. The endogeic, whose sizes range from 1 to 8 cm, have a non-pigmented body; exposure to soil with pH values below 5 restricts their activity. They live in the first few centimetres of mineral soil, feeding on humic compost and dead roots, they are capable burrowers and they make tunnels that extend mainly horizontally.

Earthworms defined as anecic, whose sizes generally are in excess of 5 cm and have a reddish-brown body, are excellent burrowers, making vertical tunnels that can reach several metres in depth; they tend to avoid soil that is asphyxiated or lacking in moisture. They live in the mineral layer of the soil but rise to the surface, mainly at night, to feed on litter. The activity of earthworms produces a significant effect, not just on the structure, but also on the chemical composition of the soil, since a large part of the organic matter ingested by earthworms is returned to the soil in a form easily used by plants. While they are feeding, earthworms also ingest large quantities of mineral substances (minimally so in the case of the epigeic), that are then mixed with the organic matter ingested and, after having been cemented with a little mucous protein, are expelled in piles called worm casts. In addition to being rich in nitrogen and other nutritive substances such as calcium, magnesium and potassium, worm casts also contain a large quantity of non-digested bacteria which proliferate easily in this sub-stratum and contribute to the humification and mineralisation of organic matter.

Vermicasts have more favourable physico-chemical properties, increased microbial population, enzyme activities and nutrient mineralization that support plant growth and yeld. Many studies have reported increased microbial activities during the passage of food through the gut in earthworms and higher numbers of fungi, bacteria and actinomycetes in vermicasts. showed enhanced microbial populations and activity in the freshly deposited pressmud vermicasts of *Lampito mauritii* and *Eudrilus eugeniae* in relation to nutrient rich substrate concentrations, multiplication of microbes after passing through the gut, optimal moisture level and large surface areas of vermicasts ideally suited for better feeding and multiplication of microbes. It is important to note that often worm casts are not released in the same layer in which the earthworm fed; in effect the components of the anecic group live at depth and release their mineral-rich faeces on the surface. In contrast the endogeic live at the surface and release their faeces, once again rich in organic matter, at lower depths.

These forms of behaviour together with the direct action due to the burrowing, ensure that the soil is mixed, thereby increasing its fertility. The burrowing of the earthworms is also essential for increasing the aeration of the soil and for improving the circulation of water since their tunnels increase the porosity of the ground by 20-30%, enabling those organisms that are not good burrowers to move around easily even at lower depths of the soil. Other than with their worm casts, earthworms contribute to the increase in the amount of nitrogen present in the ground through the excretion of ammonia and urea, forms that are directly useable by plants; furthermore a sizeable quantity of nitrogen is returned to the soil on the death of animals, which have a 72% protein content.

Within the context of edaphic fauna it is possible to identify, based on the type of locomotion, swimming organisms, capable of moving around in capillary or gravitational water, reptants that move by taking advantage of natural porosity or cavities produced by other organisms and burrowing organisms. The last-mentioned can open up cavities in the ground in various ways; for example, earthworms compress the ground outside their bodies, diplopods use their legs and backs to push the sediment upwards, the larvae of click beetles use their mandibles, while moles, scarabs and mole-crickets have specialised legs for burrowing. This continual burrowing activity contributes to the creation of spaces within the soil with the resultant increase in its porosity; the increase of the pores between the particles in turn increases the aerobic bacterial activity and the consequent speed of demolition of organic substances. This bioturbation also has positive effects on water retention, percolation processes and the development of the rhizosphere. The burrowing activity also enables the soil to be mixed and organic matter from the surface layers to be incorporated into the lower layers, while mineral substances are brought towards the surface.

Mole cricket.

The anthills is one of the most interesting and elaborate examples of the modification of the soil by cunicular burrowing organisms. It consists of a complex of chambers generally constructed on several levels and linked together by tunnels and corridors. Some nests can reach a depth of more than 5 metres and contain over 2,000 chambers, some of them set aside for the cultivation of fungi. The tunnels that connect the chambers contribute to the circulation of air and water within the anthill. Each chamber is inhabited by numerous individuals, some are reserved for incubating eggs, others for rearing the larvae, and yet others for the development of the nymphs which are moved to chambers where the humidity and temperature are more conducive to their development. An anthill can rise above ground level and can have one or more entrances, or be completely underground and communicate with the surface through one or more exits that are constantly guarded by sentries. Special devices prevent the water that penetrates the soil from flooding the chambers and in this way ensure the survival of the eggs and the nymphs, which are incapable of leaving the nest. As far as the soil is concerned, the presence of channels and tunnels increases the porosity, assisting the penetration of air and water. In addition, a consequence of the movement of the fine material towards the surface by ants during the course of the construction and maintenance of the anthill, is the creation at the surface of a layer with a fine particle size, which is more mineral than organic in nature.

A typical anthill of Formica rufa.

The soil fauna, in particularly molluscs and earthworms, also has an effect on the soil through the secretion of cutaneous mucous, that have a cementing effect on the particles in the ground, assisting the stability and structure of the soil and making it less vulnerable to processes of erosion. The mucous secretions, the faeces (especially those of earthworms) and the bodies themselves of the animals (when they die) influence in large measure the concentration of nutrients present in the soil particularly potassium, phosphorous and nitrogen, reducing the C/N ratio of the litter and facilitating decomposition.

A slug (a) and a snail of the species Helix pomatia (b).

The presence of roots is generally associated with a greater density of micro-organisms in the nearby soil compared with soil devoid of roots; the term rhizosphere is used in a broader sense to refer to the portion of soil surrounding roots in which the micro-organisms are influenced by their presence; its extension is very variable but in general it is considered to be the cylinder of soil used by the root hairs and in which they emit exudates. The rhizosphere can be distinguished from the majority of the soil on the basis of its chemical, physical and biological characteristics. Penetrating the ground, the roots act on the clay minerals and the particles of soil surrounding them; this leads to the formation of an area around them in which the water pathway, and the movement of nutrients and microflora is more heavily channelled than in the rest of the soil. For the same reasons, the organic matter released by the roots accumulates close to them. The chemical nature of the rhizosphere is significantly different from the rest of the soil; this results in large part from the release of carbon and the selective capturing of ions in solution in the groundwater by the roots. The plants act as carbon pumps fixing what is available in the atmosphere in the root exudates that are quickly captured by bacteria; for this reason the level of carbon available around the roots is never very high. The selective absorption of ions instead, causes the depletion of some of them in the rhizosphere, while others, not absorbed by the roots, tend to accumulate.

The relationship between the roots and the microflora can be very close and lead to bacteria or fungi becoming an integral part of the roots as in mycorrhizal symbiosis and in the association of bacteria and legumes. The peculiar characteristics of the rhizosphere are also reflected in a selectivity of the animal element. The interaction between soil animals and the roots of the plant can take a variety of forms that lead to benefits or repress the growth of the plant, and often involve interactions with the microbial population of the soil. The dispersion of the inocula of mycorrhizal fungi by soil animals can have beneficial effects for the plants; this dispersion is particularly favoured by burrowing organisms belonging to the mega- and mesofauna. The hyphae of mycorrhizal fungi may make up a significant proportion of the total microbial biomass in some soils and can become one of the most important sources of food for fungus grazing animals such as springtails. Numerous soil animals feed directly on the roots of plants; among them are a large number of species of springtails and myriapods. It is still not entirely clear how much of the damage inflicted on plants can be attributed to the direct action caused by the grazing of the roots or from the subsequent vulnerability of the roots to pathogens in the soil, especially fungi.

It is extremely rare for the biological relationships between soil organisms to consist of a simple and clear interaction. The actual conditions are the result of many complex interactions that typically involve multiple participants in life in the soil such as the plants, the microbes, the fungi and the animals, the last mentioned at different levels of organization. For example, examination of the faeces of earthworms reveals a mixture of fragments of plants, microorganisms, fungi, and fragments of encysted animals and protozoa capable of surviving the unfavourable conditions in the gut of the earthworm.

Diversity of Soil Fauna in Different Ecosystems and Soil Managements

A rapid survey of invertebrate and vertebrate groups reveals that at least ¼ of described living species are strictly soil or litter dwelling. The greatest diversity is observed in systems where equilibrium exists between the productivity level and the perturbation rate, whereas local species extinctions may occur through lack of demographic recuperation when perturbations increase or through competitive exclusion when productivity increases. Land use changes and the intensification of agriculture also generate severe habitat degradation or destruction of soil biota. The environment of soil organisms in managed ecosystems can be influenced by any combination of land use factors, such as tillage, pesticide and fertilizer application, soil compaction during harvest, and removal of plant biomass.

The profile of a mature and undisturbed soil can be divided into a number of spheres, whose chemico-physical properties are determined by vegetation cover, by geographical location and by climate; this stratification is reflected in the edaphic population and as a result in its vertical distribution. Usually two groups of organisms are recognised in the soil: the euèdaphons, which include organisms that live in the strip of mineral soil, and the hemiedaphons, found in the strip of organic soil; to these can be added the epiedaphons (or epigeons), consisting of organisms that live in the surface of the soil, and hyperedaphons that extend to the herbaceous layer. The moisture content and the pH of the soil have the greatest influence on the distribution of the hemi- and euedaphon fauna, even if the characteristics of the litter, the porosity of the soil and numerous other factors are also important in determining the vertical distributions of edaphons. Humidity has often been used to establish further subdivisions within the hemiedaphons; one example is the classification in which hemiedaphon springtails were subdivided into three categories: hydrophilic (living on the surface of free water), mesophilic (living in damp organic litter) and xerophilic (living in more exposed dry areas, such as in lichens, mosses and tree bark).

In reality this type of stratigraphical classification is difficult to apply since soil organisms migrate on both a daily and a seasonal basis; in fact, many species, for example mites, springtails and isopods, can move towards the surface over distances ranging from a few millimetres to a few centimetres. An interesting case is that of sinfili that can go down as much as 40-50 cm. Nematodes are usually most abundant in the first 10 cm of soil, but their distribution can vary greatly depending on the type of vegetation cover and conditions of dampness. For example, in soil with abundant vegetation cover and a high presence of roots, these organisms concentrate in the upper part of the ground which, being rich in exudates and decomposing organic matter, maintain a numerous population of their potential prey. In soils with sparse vegetation cover, the high frequency of dry conditions on the surface cause nematodes to be more common at a depth of 5-10 cm whereas, in cultivated land a uniform distribution of the population can be observed down to a depth of about 20 cm, attributable to the mixing effects caused by ploughing. The density of these organisms is particularly high in grasslands, where they can be present with a density in the order of 20 million/m^2, whereas in moorland the density varies greatly and usually is less than one million/m^2, even though in some cases it can exceed 5 milioni/m^2.

The abundance, diversity, composition and activity of species of the soil community can be affected by plant species, plant diversity and composition, as well as by animal grazing. The two factors that most likely influence the soil community are the nutrient resources available, and the diversity of microhabitats. Resource type is determined by the tissue chemistry of plant species and the nutrient content of excreted wastes from grazing by consumers. The microhabitat is directly affected by the physical and chemical properties of the system as well as the amount of nutrient input.

Various studies have described the structure of soil invertebrate communities in relation to forest diversity, dynamics and management. In central Italy silvicultural practices and the composition of deciduous forests do not seem to have any important effect on the structure of microarthropod communities. The absence of a change in soil community structure could be linked to the litter layer that in these hardwood stands is thick enough to maintain a high level of organic matter and a favourable microclimate in every season. Soil mesofauna seem to recover quickly after disturbances such as tree cutting indicating a good level of ecosystem integrity (community resilience). The same aspect also emerged in conventional tillage conditions where soil arthropod abundance was significantly higher in autumn compared with summer. The study suggested that, given sufficient time without soil disturbance, soil arthropods are able to recover within the growing season. These results agree with studies conducted on a temperate cool rain forest in west Canada where there were no significant differences in the population density of arthropods between undisturbed forests on harvested plots and those on unharvested patches.

Additionally, in another recent study involving a beech forest, the hypothesis of, which predicts community changes during forest rotations, was refuted from a functional view point. However several studies considering the effects of silvicultural practices on soil fauna found important impacts on soil forest fertility/productivity and in the terrestrial food chain. It is generally accepted that the removal of trees by clear-cutting, or other methods, has a significant effect on the invertebrate fauna of the forest floor. The effects on arthropod communities are complex and difficult to analyze since various taxonomic groups are affected and they react to impacts differently.

A study that compared the soil community in different grasslands with a semi-natural woodland area and an arable land site showed that microarthropod communities of the three land use typologies differed in terms of both observed groups and their abundance. Typical steady soil taxa characterised woodland and grassland soils, whereas their abundances were significantly higher in the former. The mean highest abundances of Acari and Collembola were observed in grassland, confirming the suitable trophic conditions of this habitat, whose taxa diversity and soil biological quality did not significantly differ from woodland samples. On the contrary in the arable land, the microarthropod community showed a reduction both in taxa number and soil biological quality compared with the other sites. Besides, mite and springtail abundances were significantly lower than in grassland. The study concluded that soil biological quality and edaphic community composition highlighted the importance of grassland habitat in the protection of soil biodiversity, especially because it combines fauna conservation with the production of resources for human needs.

Principal Factors Affecting the Loss of Soil Fauna Diversity

Around the world there are numerous soils that have lost their fertility or their capacity to carry out their function due to the impact of man. The causes are mainly related to processes that are accelerated or triggered directly by human activities and that often act in synergy with each other, amplifying the effect. Among them, the most widespread at a worldwide level are erosion, the loss of fertility and a decline in organic matter, compaction, salinisation, phenomena of flooding and landslides, contamination and the reduction in biodiversity. Reduction of soil biodiversity as a result of urbanization can be even more severe. The urbanization process leads to the conversion of indigenous habitat to various forms of anthropogenic land use, the fragmentation and isolation of areas of indigenous habitat, and an increase in local human population density. The urbanization process has been identified as one of the leading causes of declines in arthropod diversity and abundance.

Soil properties determine ecosystem function and vegetation composition/structure, serve as a medium for root development, and provide moisture and nutrients for plant growth. Disturbances linked to natural forces and to human activities can alter physical, chemical and biological properties of soils, which can, in turn, impact long-term productivity. Humans have extensively altered the global environment and caused a reduction of biodiversity.

These change in biodiversity alter ecosystem processes and change the resilience of ecosystems to environmental change. It is estimated that human activities increased the rates of extinction 100-1000 times. In the absence of major changes in policy and human behaviour our effect on the environment will continue to alter biodiversity. Land use change is projected to have the largest impact on biodiversity by the year 2100. Within agricultural land use, that covers 10.6 % of land surface, the intensity of agronomic practices and crop management can also affect biodiversity. Land use is considered to be the main element of global change for the near future. Forests, tropical or temperate, generally represent the biomes with the largest soil biodiversity. Consequently any land use change resulting in the removal of perennial tree vegetation will produce a reduction of soil biodiversity. In some cases forests are succeeded by pasture or perennial grasslands, while in others arable land replaces formerly wooded areas. The change in soil biodiversity will therefore be influenced by the subsequent use of land following the forest.

Val D'Adige (Italy), where it's possible to note the intense agricultural utilization.

The abundance, biomass and diversity of soil and litter animals are influenced by a wide range of management practices which are used in agriculture. These management practices include variations in tillage, treatment of pasture and crop residues, crop rotation, applications of pesticides, fertilisers, manure, sewage and ameliorants such as clay and lime, drainage and irrigation, and vehicle traffic. Furthermore differences in agricultural production systems, such as integrated, organic or conventional systems, have been demonstrated to affect soil fauna in terms of numbers and composition. The impact of soil tillage operations on soil organisms is highly variable, depending on the tillage system adopted and on soil characteristics.

Conventional tillage by ploughing inverts and breaks up the soil, destroys soil structure and buries crop residues determining the highest impact on soil fauna; the intensity of these impacts is generally correlated to soil tillage depth. Minimum tillage systems can be characterized by a reduced tillage area (i.e. strip tillage) and/or reduced depth (i.e. rotary tiller, harrow, hoe); crop residues are generally incorporated into the soil instead of being buried. The negative impact of these conservation practices on soil fauna is reduced compared with conventional tillage. Under no-tillage crop production, the soil remains relatively undisturbed and plant litter decomposes at the soil surface, much like in natural soil ecosystems. The influence on soil organism populations is expected to be most evident when conservation practices such as no-till are implemented on previously conventionally tilled areas because the relocation of crop residues to the surface in no-till systems will affect the soil decomposer communities. No-till and minimum tillage generally determine an increase in micro-arthropod numbers.

Conventional tillage by ploughing inverts and breaks up the soil, destroy soil structure and buries crop residues.

A multidisciplinary study was carried out over four years in Northern Italy on a silt loam under continuous maize to evaluate two factors: the soil management system (conventional tillage and no-tillage) and N fertilisation. This study showed that total micro-arthropod abundances were higher in NT compared with CT (+29%). Acari showed higher sensitivity to tillage compared with collembola. Moreover, N fertilisation with 300 kg N ha^{-1} had a negative effect on the total micro-arthropod abundance. Also observed that higher values of mite density were associated with a decrease in tillage impact. Similarly reported that the mite community, in particular oribatid, was more abundant in no-tillage compared with conventional tillage, but the differences were found only in some periods of the year. Conventional tillage caused a reduction of micro-arthropod numbers as a result of exposure to desiccation, destruction of habitat and disruption of access to food sources. The influence of these impacts on the abundance of soil organisms will be either moderated or intensified depending on their spatial location; that is, in-row where plants are growing, near the row where residues accumulate or between rows being subjected to possible compaction from mechanized traffic.

Observations on the impacts of different forms of agricultural management on communities of microarthropods showed that the high input of intensively managed systems tends to promote low diversity while lower input systems conserve diversity. It is also evident that high input systems favour bacterial-pathways of decomposition, dominated by labile substrates and opportunistic, bacterial-feeding fauna. In contrast, low input systems favour fungal pathways with a more heterogeneous habitat and resource leading to domination by more persistent fungal feeding fauna.

The effects of fertilizers on soil invertebrates are a consequence of their effects on the vegetation and, directly on the organisms. Increases in quantity and quality of food supplied by vegetation are frequently reflected in greater fecundity, faster development and increased production and turnover of invertebrate herbivores. The effects of organic and inorganic fertilizers in terms of nutrient enrichment may be comparable, but these two types of fertilizers differ in that organic forms provide additional food material for the decomposer community. Concluded that the total soil microbial biomass and the biomass of many specific groups of soil organisms will reflect the level of soil organic matter inputs. Hence, organic or traditional farming practices, that include regular inputs of organic matter in their rotation, determine larger soil communities than conventional farming practices. Also reported that the soil microbial and faunal feeding activity responded to the application of compost with higher activity rates than with mineral fertilization. Generally the responses of soil fauna to organic manure will depend on the manure characteristics, and the rates and frequency of application. Herbivore dung, a rich source of energy and nutrients, is exploited initially by a few species of coprophagous dung flies and beetles and, later, by an increasingly complex community comprising many general litter-dwelling species.

A study related to sewage sludge application on agricultural soils showed an increase in the abundance of Collembola, Carabidae, Oligochaeta, soil nematodes and Arachnida. In some cases, the application of sewage sludge to agricultural land can bring toxic substances that, accumulating in the soil, reach potentially toxic levels for soil fauna. Field studies have suggested that metals contained in sewage sludge don't reduce the abundance of euedaphic and epigeic collembola but may alter their population structure. Reported negative effects on collembola communities in soil treated with sewage sludge and this effect may be attributed to anaerobic conditions and high ammoniacal level. In effect, the knowledge gained in relation to the effects of sewage sludges showed that species more sensitive to the toxic substances contained in sewage sludge can disappear, while others which are more tollerant, can dramatically increase.

Organic wastes are usually stored in dump areas with associated high management costs and problems of environmental impact. On the other hand organic wastes could become an easily available and cheap source of organic matter after composting processes. The use of compost obtained by organic waste in agricultural activity enables the converting of waste materials into a useful resource. Therefore, the national authorities have over recent years stimulated both the use of compost to reduce the soil fertility loss and research aimed at assessing its effects on both agricultural production and soil environment. But negative effects on soil fauna could be related to the use of organic waste, such as the accumulation of trace metals in soil. In fact many trace elements contained in organic wastes can reduce the abundance and diversity of soil microarthropod communities, or can influence the survival potential and the rate of growth of more sensitive species. concluded that negative effects of compost use on soil fauna abundance or biodiversity were not observed in two Italian sites studied that were treated with compost, supporting the use of this product derived from waste in order to add organic matter into the soil.

Pesticide application to the soil can affect the soil fauna influencing the performance of individuals and modifying ecological interactions between species. When one or more ecosystem component is impacted by pesticides, this will affect the microarthropod communities in terms of number and composition. Pesticide toxicity on soil fauna is determined by different factors, such as the pesticide's chemical and physical characteristics, the species' sensitivity and the soil type. In fact among soil microarthropods, different taxa showed a variety of responses. The physical and chemical characteristics of the soil, such as its texture, structure, pH, organic matter content and quality, and nature of clay minerals, are important factors determining the toxic effects of pesticides or other xenobiotics. A study carried out by showed reduced toxic effects, as a function of soil type, in the following order: sand>sandy-loam>clay>organic soil. Often, the toxicity of pesticides can be related directly to soil organic matter content. However pesticide application does not always cause negative impacts on the entire soil microarthropod community. For example for certain types of soil there is evidence that some taxa can obtain a competitive advantage from the application of some specific pesticides.

Use of Soil Invertebrates in Soil Biodiversity Assessment and as Soil Biological Quality Indices

The increasing anthropic pressure on the environment is leading, in most parts of the world, to a rapid change in land use and an intensification of agricultural activities. These processes often result in soil degradation and consequently loss of soil quality. Soil quality could be defined as the capacity of a specific kind of soil to function, within natural or managed ecosystem boundaries, to sustain plant and animal productivity, to maintain or enhance water and air quality, and to support human health and habitation. A common criterion for evaluating the long-term sustainability of ecosystems is to assess the fluctuations of soil quality. Soil reflects ecosystem metabolism; within soils, all bio-geo-chemical processes of the different ecosystem components are combined. Monitoring ecosystem components plays a key role in acquiring basic data to assess the impact of land management systems and to plan resource conservation. Maintaining soil quality is of the utmost importance to preserving biodiversity and to the sustainable management of renewable resources.

Soil quality can be evaluated through its chemical-physical properties and biological indicators and indices. The importance of some of these parameters is generally accepted. Soil organic matter among the chemical indicators, bulk density and aggregate stability among the physical indicators, were the most often used but there were few examples of biological indicators of soil quality. However, biological monitoring is required to correctly assess soil degradation and correlated risks. Indicators of soil health or quality should fulfil the following criteria: 1) sensitivity to variations of soil management; 2) good correlation with the beneficial soil functions; 3) helpfulness in revealing ecosystem processes; 4) comprehensibility and utility for land managers; 5) cheap and easy to measure. The growing interest in the use of living organisms for the evaluation of soil conditions is justified by the great potential of these techniques, that allow the measurement of factors difficult to detect with physical-chemical methods and give more easily interpretable information. Biotic indices, based on invertebrate community studies, were recently developed as a promising tool in soil quality monitoring. These organisms are highly sensitive to natural and human disturbances and are increasingly being recognized as a useful tool for assessing soil quality.

The complex relationships of soil fauna with their ecological niches in the soil, their limited mobility and their lack of capacity to leave the soil environment, make some taxa (e.g. Collebola, Protura, Pauropoda) particularly vulnerable to soil impact. For these reasons soil fauna communities represent an excellent candidate for soil bioindication and for evaluating soil impact. The basic idea of bio-indications is that the relationship between soil factors and soil communities can be tight. When soil factors influence community structure, the structure of a community must contain information on the soil factor. To retrieve information about soil quality, different properties of community structures, such as the richness and diversity of species, the distribution of numbers over species, the distribution of body-size over species, the classification of species according to life-history attributes or eco-physiological preferences and the structure of the food-web can be used. The number of bio-indicator systems using soil invertebrates is relatively high; some approaches use Nematode, Enchytraeid, mites, Collembola, Diptera, Coleoptera or all microarthropod communities. Moreover the use of bioindicators makes it possible to highlight the interactions among the different pollutants and between them and the soil. Often, bio-monitoring techniques are not very specific in identifying the pollutant or environmental variable that creates stress in the organisms. For this reason bio-monitoring must not be considered a substitute for physical-chemical analysis.

Most edaphic animals have life cycles that are highly dependent on their immediate environment, interacting with soil in several different ways. To be able to evaluate their role and function, it is important to use methodologies that highlight either the number of species present or the processes and roles that they play in the soil environment. In particular mesofauna groups are a key component of soil biota. They are very abundant, their role in soil formation and transformation is well-recognized, the area covered during their life cycle is representative of the site under examination, their life histories permit insights into soil ecological conditions and, several species have already been recognized as useful biological indicators of soil quality. In general, soil invertebrate-based indices consider the consistency and richness of populations. Some species in a single taxon may be specified as indicators of soil quality or as test organisms and used in toxicology tests. In the collembolan taxon, *Folsomia candida* is the most frequently used species in both sub-lethal and lethal testing. *Onychiurus armatus, Orchesella cincta, Isotoma notabilis, Tetrodontophora bielanensis* and other collembolan species have been used in laboratory tests but have not reached the same level of routine use as has *F. candida*. Because of the species-specific differences in responses to contaminants, the tests conducted on *F. candida* provide partial indications as to the effects provoked by these substances on the collembolans; this information has also been useful in calibrating experiments on other species. Some collembolan species like *Folsomia quadrioculata, Folsomia fimetariodes, Isotoma minor* and others species have been used to evaluate the effects of chemicals on collembola in the field.

Folsomia candida, a collembola used in toxicology tests.

The Maturity Index (MI) is a bio-indicative method based on the soil nematological community composition that sorts the families into five categories according to the reproductive characteristics which define them as colonizer or persistent organisms. Each family is assigned a score ranging between 1 and 5 passing from the colonizer to persistent forms. These values, called c-p (v), are then multiplied by the organisms' frequency (f) and finally inserted in a summation.

$$MI = \sum_{i=1}^{n} v(i) f(i)$$

Index values close to 1 indicate the predominance of colonizer forms and therefore an environmental situation that presents no great stability. On the other hand, values ranging between 2 and 4 highlight the presence of a situation with persistent forms and more stable conditions.

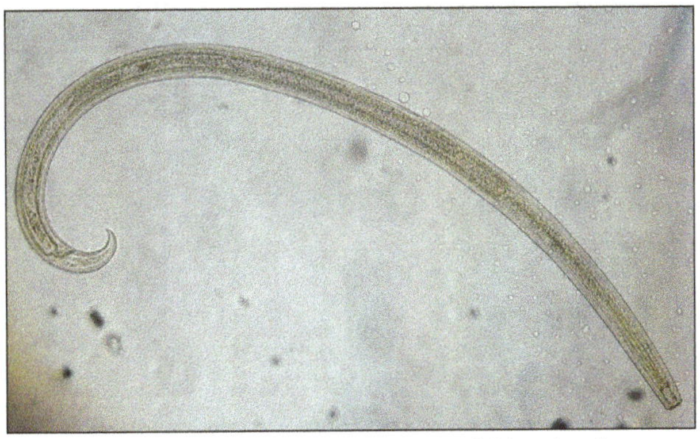

A nematode belonging to Mononchidae family.

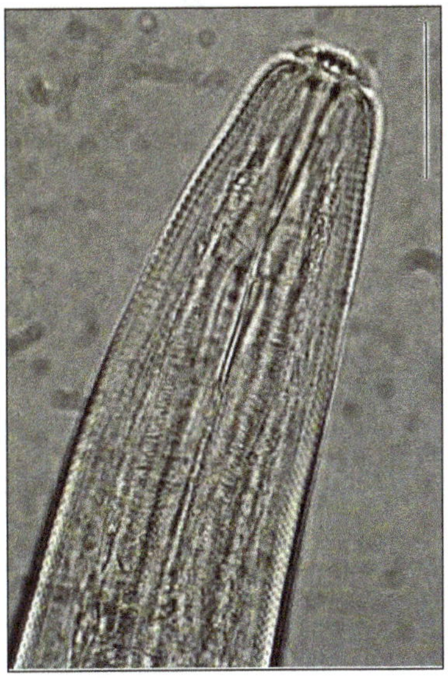

A nematode belonging to Plectidae family (Chiloplectus andrassyi). Its head is shown in the picture.

Moreover, soil biological quality could be expressed using an Acari/Collembola ratio (A/C) and the QBS-ar index. The first indicator is based on the densities of Acari and Collembola communities where in natural conditions the ratio of the number of mites to the number of collembola is greater than one. On the other hand, in the case of soil degradation, that ratio shifts towards collembola and its value diminishes. The QBS-ar index is based on the following concept: the higher the soil quality is, the higher the number of microarthropod groups morphologically well adapted to soil habitat will be. Soil organisms are separated into biological forms according to their morphological adaptation to the soil environment; each of these forms is associated with a score named EMI (eco-morphological index), which ranges from 1 to 20 in proportion to the degree of adaptation. The QBS-ar index value is obtained by summing the EMI of all the collected groups. If biological forms with different EMI scores are present in a group, the higher value (more adapted to soil form) is selected to represent the group in the QBS-ar calculation. QBS-ar was applied in several agricultural ecosystems, grasslands, urban soil and woods at different levels of naturality and anthropic impact. This index reached higher values in grassland and woods and lower values in agricultural ecosystems. Moreover, demonstrated that QBS-ar was highest in the soil with the lowest metal content and the highest density and taxa richness of the invertebrate community. The study suggested that this index seems to be appropriate in defining the quality of the investigated soils. In Figure are showed some QBS-ar values detected in different soils.

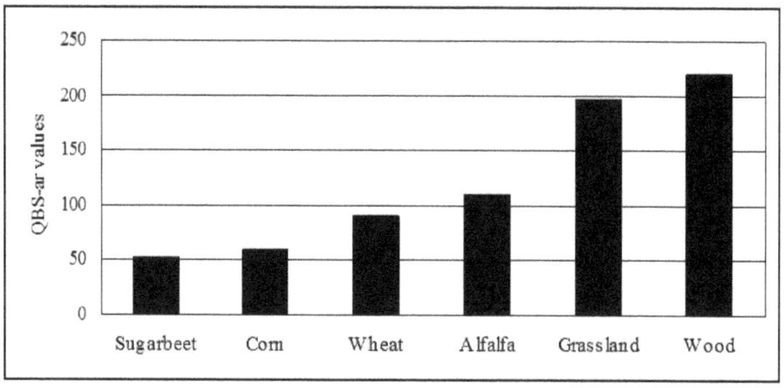

QBS-ar values detected in different soils.

Forest ecosystem management, QBS-ar could be an efficient index for evaluating the impacts on soil of forest harvesting (i.e. soil compaction due to logging) and so for determining the sustainable use of renewable resources. At the same time, QBS-ar can be a valuable tool in ecosystem restoration programmes for monitoring the development of soil functions and biodiversity and for preventing the negative effects of soil compaction when mechanization is used (e.g. in Europe many LIFE projects include mechanized operations). Furthermore this index could be implemented in environmental management programmes of urban forestry and protected areas in relation to recreational use to prevent the negative effects of trampling.

Another index that could be applied to soil fauna communities is the V index, which expresses the magnitude of the response to tillage. The V index was calculated as:

$$V = \frac{2M_{CT}}{M_{CT} + M_{NT}} - 1$$

where M_{CT} and M_{NT} could be the abundance of taxa under conventional tillage and no-tillage, respectively. Six magnitude categories were provided for the V index:

- Extreme inhibition by tillage (or treatment): $V < -0.67$.
- Moderate inhibition by tillage (or treatment): $-0.33 > V > -0.67$.
- Mild inhibition by tillage (or treatment): $0 > V > -0.33$.
- Mild stimulation by tillage (or treatment): $0 < V < 0.33$.
- Moderate stimulation by tillage (or treatment): $0.33 < V < 0.67$.
- Extreme stimulation by tillage (or treatment): $V > 0.67$.
- Wardle V index proved to be a good indicator of the response to tillage.

Effect of Actions for Restoration and Conservation of Soil Fauna Diversity

The realisation that degradation of the soil is an environmental problem of global significance, with immediate consequences at an economic and social level, and the recognition of the importance of protecting it, has led to an increase in international initiatives. The Convention on Biological Diversity is the first global agreement aimed at conservation and the sustainable use of biological diversity. The CBD lies at the heart of biodiversity conservation initiatives. It offers opportunities to address global issues at a national level through locally grown solutions and measures. One important requirement is the development of National Biodiversity Strategies and Action Plans mainstreaming them into relevant sectors and programmes, a principal means for implementation of the Convention at the national level.

The recent Conference of the Parties of the Convention on Biological Diversity demonstrated that the need for action to protect biodiversity is unanimously acknowledged. Biodiversity conservation is essential both for ethical reasons and especially for the ecosystem services that the complex of living organisms provides for current and future generations. These ecosystem services are essential for the functioning of our planet. A necessary starting point for achieving the objective of preserving soil biodiversity is to reach an adequate level of knowledge on its extent and on its spatial and temporal distribution. Among the most important tasks that man should set himself for safeguarding the assets of the soil is to rectify the damage caused to ecosystems, where it is still possible, through works of environmental recovery. There are a number of actions that could reduce the risk of damage to ecosystems, ecological receptors and to humans that include land reclamation, environmental restoration and halting the exposure to sources of pollution (either by physical means or through communication such as information and education).

These initiatives are often problematical and carry a heavy price tag. It is well known that the reclamation of waste disposal sites is usually characterized by soil quality problems; soil used to cover the dump is generally affected by physical, biological and sometimes chemical degradation. These conditions affect both the plant and the animal communities and, more generally, the effectiveness of the restoration processes. In this habitat the soil fauna can be inherited or can have established itself on the reclaimed waste disposal site; generally the more structured the soil is the more complex is the soil fauna community.

Management of waste disposal is an extremely emotive issue and the level of acceptance of this kind of facility by the local community is often dependent on the mitigation of its impact and on a good restoration of the interested sites. Beyond the well-known hygienic-health risks and its strong impact on the landscape, the construction of a dump requires the removal of a large quantity of soil that cannot be considered an unlimited resource. In order to prevent the risks of groundwater contamination, the bottom of waste disposal sites is sealed using layers of waterproof materials and the top of the dump is sealed to prevent rain seepage and the leakage of biogas produced by decomposition. The surface layer of a dump is usually covered by filling soil that is suited to being re-colonized by plants and animals, even though this soil does not represent the same physical, chemical and biological characteristics of the removed soil. Thinking about the complexity of the issues relating to dumps, it is important to obtain exhaustive and multi-disciplinary information on the environment and it is necessary to support traditional physical-chemical analyses with bio-monitoring techniques.

The growing interest in the employment of living organisms for the evaluation of soil conditions is justified by the great potential of these techniques, that allow the measurement of factors difficult to detect with physical-chemical methods and that give more easily interpretable information. The great differences in abundance and maturity shown by nematode communities (Maturity Index MI 2.76) in the top soil of a reclaimed waste disposal site compared with permanent grassland and wood could have been caused by many factors. In this study disturbance in the soil from a dump is reflected by the lower maturity due to the absence of omnivorous nematodes like Thornenematidae.

The greater abundance and maturity in the nematode communities from woods and grasslands may be due to the fact that the soil is less disturbed. The differences in soil litter composition and in root distribution had probably affected the predominance of plant feeding nematodes in grasslands and the predominance of dorylaimids (Qudsianematidae, Leptonchidae) in the woods, in a study related to nematode communities in ash dumps covered with turf and reclaimed from different times reported MI values ranging from 2.0 and 2.3. reported that in the ash dump reclaimed over a longer period the total abundance of nematodes was higher than those reclaimed over a shorter time and in some samples it was similar to the lowest abundances observed in grasslands in Poland. The author suggested that the species with high ability to colonize new habitats had the best chance of survival in these conditions. Probably, lack of soil structure, high salt content and low organic matter content may be responsible for low MI values and the low density observed in the dump that was the subject of the study. Moreover, the poor and little structured covering of vegetation in the dump, that consequently did not create homogeneity in soil structure and organic matter content, may be a very important reason also limiting the microarthropod community showed the vulnerability of springtails and pauropods.

The study observed that the reduction of collembolan and pauropod densities in high-input management systems is largely explained by the mechanical and chemical perturbations produced by conventional agricultural management practices and by particular abiotic soil conditions present in the intensively managed sites that are unfavourable for these organisms. The study reported that symphylans were more abundant in the mixed management site. Extraction activities have a significant impact on the community, affecting both vegetation and soil microbes and animals. The studies related soil community changes during ecological succession in degraded soils are still scarce.

After the extraction, the ecosystem would be able to recover spontaneously if the mineral substratum and the environmental conditions were right, but in many cases the physical, chemical and biological conditions of the soil are too disturbed (e.g. unbalanced particle sizes, low organic matter content, inadequate biological component of the soil) or the start of a secondary succession is impeded due to isolation from the colonisation resources. In the process of open-cast mining, the vegetation is completely removed and this causes major changes in the physical, chemical and microbiological properties of the soil. Topsoil is an essential component in abandoned quarries for the growth of vegetation and must be preserved for the restoration of the ground once the extraction work has been completed. Generally a significant period of time passes between the initial removal of the topsoil and the final distribution of the same over the restored area. Because of this, the properties of the stored soil can deteriorate and it can become biologically sterile. In it was demonstrated that the microbial population in the accumulated stockpile falls dramatically in comparison with a control sample of soil that had not been removed.

In the same study results were compared of samples taken in the quarry and those taken from a control area, and the particle sizes of the mineral components were analysed. It was found that the proportion of sand particles in the quarry site had risen, while the particles of lime and clay had fallen in comparison with the control soil, phenomena probably due to the process of erosion. This is a consequence of a low stability of the aggregates and, consequently, a high rate of infiltration. In the accumulated stockpile instead, it was observed that there was an increase in density and a reduction of porosity, caused by compaction by machinery during the excavation. These changes make the diffusion of gases more difficult and they restrict the growth of the deep roots of the plants, thus representing one of the reasons why in the shrub stage they cease to grow. To this must be added a change in the pH of the stockpile, with an increase in its acidity due to the separation of the base cations and the scarcity of nutrients, probably caused by the reduction of soil microbes induced by the accumulation of the soil stockpile.

If it is not possible to deposit the stored soil in the quarry site within the maximum period for preserving its fertility, it becomes necessary to initiate a biological restoration in order to preserve the topsoil, but it must still be carried out within the conservation period, that is before the cessation of microbiological activity and the breakdown of the nutrient cycle, in order to prevent the soil from becoming completely unproductive. The motor for the succession is the interaction between the trophic levels of the ecosystem. Given that plants are at the bottom of the food chain and that they play an important role in the formation of the physical structure of the soil, the changes in the vegetation during the succession are crucial to the successional development of the other organisms, including soil animals.

At the same time, the succession of plants depends on the abiotic conditions of the site, on the pool of species and the intra-species competition, but it is also influenced by other trophic levels, among them the herbivores and the soil invertebrates. The latter can influence the successional changes of the plants through soil phytophagous, the effects on the availability of nutrients, and by influencing the formation and the modification of the soil as a habitat for plants. Soil communities are important in the processes of soil formation because they influence the distribution of the organic matter and as a consequence, the rate of decomposition. The restoration of an extraction site demonstrated that there are strict timing synchronisations between the changes in the vegetation, the soil and its being populated. This indicated that the interaction between all these components can play an important role in successional changes in the ecosystem. The study of these components of fauna are therefore important for monitoring environmental recovery processes, given the links between edaphic fauna, soil and vegetation.

As with vegetation, the post-restoration recovery of the invertebrate community is slow and not less than 15 years, with 80-102 years estimated for the recovery of springtail communities in forests. The maximum density of the mesofauna is generally reached during the 2-3 years of the "pioneering" phase, which is followed by a drastic reduction in the density to levels of less than 20% within the following 10 years. Successive changes in the taxonomical composition and relative abundance may be correlated to successional changes in factors such as vegetation cover, the pH of the soil and the content of organic matter, etc. The richness of the animal taxa is indicative of the maturity of the community of vegetation in the recovered area. After recovery of the soil, the process of secondary succession involves an increase in the diversity of the structure and in the available ecosystem energy that facilitate the development of high trophic levels.

A study conducted in northern Italy in an open-cast quarry after the restoration phase showed mature microarthropod communities and higher abundances in the sites where the extraction activity had finished earlier. The presence of edaphic organisms generally associated with stable soil conditions, such as pauropods, symphylans, proturans and diplurans, was found only in these sites (personal data unpublished). Succession to naturalized grassland from former agricultural land and pasture is accompanied by changes in plant biodiversity and in the soil community. These changes are the result of a reduction or elimination of management, fertilizer applications and of grazing by large herbivores. The response of the soil faunal community and diversity might not be in-step with plant succession.

Species of soil biota found in early successional stages persist in later stages, although with changes in dominance and species frequency. Species replacement is either less pronounced or it occurs on a different time scale. In a study of grassland succession from 7 to 29 years into restoration, the changes in soil faunal species in Isopoda, Chilopoda and Diplopoda did not correspond with plant successional changes, although the macro-invertebrate diversity and density increased with field age, but decreased in the oldest field. Environmental changes during succession increased the amount of basal resources that provided various micro-habitat and nutrient resources for macro-invertebrates that could lead to the establishment of a diverse community. An increase in the amount of habitable space created by increasing pore surface area would increase the abundance of the macro-invertebrates. It is still not clear to what extent different groups of organisms, such as nematodes, microarthropods or bacteria, respond separately or as an integrated food web community to plant succession. Mechanisms of feedback interactions of soil organisms among themselves and with roots are complex, and not well understood at a molecular level.

Soil Erosion

Soil erosion is an important social and economic problem and an essential factor in assessing ecosystem health and function. Soil erosion is one of the naturally occurring problems in soils. It will affect all landforms. Soil erosion may also happen through forces associated with farming activities such as tillage. Topsoil, which is rich in organic matter, high fertility and soil life, is relocated elsewhere "on-site" where it builds up over time or is carried "off-site" where it fills in drainage channels. Soil erosion reduces cropland productivity. Soil erosion contributes to the pollution of adjacent watercourses, wetlands and lakes.

Soil erosion refers to the wearing away of a field's topsoil by the natural physical forces of water and wind. It can be a slow process. It is relatively unnoticed or can occur at an alarming rate, causing serious loss of topsoil.

Soil compaction, low organic matter, loss of soil structure, poor internal drainage, salinisation and soil acidity problems are other serious soil degradation conditions that can accelerate the soil erosion process.

Soil Erosion, whether it is by water, wind or tillage, involves three distinct actions – soil detachment, movement and deposition. Accelerated erosion is largely the consequence of human activity. The primary causes are tillage, grazing, and cutting of timber. The rate of erosion can be increased by activities other than those of humans. Fire that destroys vegetation and triggers erosion has the same effect.

Global Problem

Erosion is a major problem affecting soils all over the world. The rapid growth of the world's population has resulted in increased cultivation of land. This puts more pressure on land and leads to soil losing its structure and cohesion, which means that it can be eroded more easily.

Heavy farming machinery can also 'compact' soil, which causes water to run straight off the surface after rain, taking soil particles with it, instead of infiltrating into the soil. The total land area subjected to human-induced soil degradation is estimated at about 2 billion ha. Of this, the land area affected by soil degradation due to erosion is estimated at 1100 Mha by water erosion and 550 Mha by wind erosion.

Gravity Erosion

Mass-Wasting is the down-slope movement of rock and sediments, mainly due to the force of gravity. Mass-wasting is an important part of the erosional process, as it moves material from higher elevations to lower elevations where transporting agents like streams and glaciers can then pick up the material and move it to even lower elevations. Mass-wasting processes are occurring continuously on all slopes; some mass-wasting processes act very slowly.

Slumping happens on steep hillsides, occurring along distinct fracture zones, often within materials like clay that once released, may move quite rapidly downhill. Surface creep is the slow movement of soil and rock debris by gravity which is usually not perceptible except through extended observation.

Water-Borne Soil Erosion

The rate and magnitude of soil erosion by water is controlled by the following factors:

- Rainfall and Runoff
- Soil Erodibility
- Slope Gradient and Length
- Cropping and Vegetation
- Tillage Practices

Rainfall and Runoff

The greater the intensity and duration of a rainstorm, the higher the erosion potential. The impact of raindrops on the soil surface can break down soil aggregates and disperse the aggregate material. Lighter aggregate materials such as very fine sand, silt, clay and organic matter are easily removed by the raindrop splash and runoff water. Soil movement by rainfall (raindrop splash) is usually greatest and most noticeable during short-duration, high-intensity thunderstorms. Surface water runoff occurs whenever there is excess water on a slope that cannot be absorbed into the soil. Reduced infiltration due to soil compaction, crusting or freezing increases the surface runoff and soil erosion. Runoff from agricultural land is greatest when compared with other land areas.

Soil Erodibility

Soil erodibility – susceptibility of soil to agent of erosion - is determined by inherent soil properties e.g., texture, structure, soil organic matter content, clay minerals, exchangeable cations and water retention and transmission properties. Climatic erosivity includes drop size distribution and intensity of rain, amount and frequency of rainfall, run-off amount and velocity, and wind velocity. Important terrain characteristics for studying soil erosion are slope gradient, length, aspect and shape. Ground cover exerts a strong moderating impact on dissipating the energy supplied by agents of soil erosion. Soil erodibility is an estimate of the ability of soils to resist erosion, based on the physical characteristics of each soil.

Texture is the principal characteristic affecting erodibility, but structure, organic matter and permeability also contribute. Generally, soils with faster infiltration rates, higher levels of organic matter and improved soil structure have a greater resistance to erosion.

Sand, sandy loam and loam-textured soils tend to be less erodible than silt, very fine sand and certain clay-textured soils. Tillage and cropping practices that reduce soil organic matter levels, cause poor soil structure, or result in soil compaction, contribute to increases in soil erodibility. The formation of a soil crust, which tends to "seal" the surface, also decreases infiltration.

Slope Gradient and Length

The steeper and longer the slope of a field, the higher the risk for erosion. Soil erosion by water increases as the slope length increases due to the greater accumulation of runoff.

Consolidation of small fields into larger ones often results in longer slope lengths with increased erosion potential, due to increased velocity of water, which permits a greater degree of scouring (carrying capacity for sediment).

Cropping and Vegetation

The potential for soil erosion increases if the soil has no or very little vegetative cover of plants and/or crop residues. Plant and residue cover protects the soil from raindrop impact and splash, tends to slow down the movement of runoff water and allows excess surface water to infiltrate.

The erosion-reducing effectiveness of plant and/or crop residues depends on the type, extent and quantity of cover. The effectiveness of any protective cover also depends on how much protection is available at various periods during the year, relative to the amount of erosive rainfall that falls during these periods.

Crops that provide a full protective cover for a major portion of the year (e.g., alfalfa or winter cover crops) can reduce erosion much more than can crops that leave the soil bare for a longer period of time (e.g., row crops), particularly during periods of highly erosive rainfall such as spring and summer.

Tillage Practices

The potential for soil erosion by water is affected by tillage operations, depending on the depth, direction and timing of plowing, the type of tillage equipment and the number of passes.

Minimum till or no-till practices are effective in reducing soil erosion by water. Tillage and other practices performed up and down field slopes creates pathways for surface water runoff and can accelerate the soil erosion process.

Forms of Water Erosion

The major forms of water-borne soil erosion are:

- Sheet Erosion
- Rill Erosion
- Gully Erosion
- Bank Erosion

Sheet Erosion

Sheet erosion is the movement of soil from raindrop splash and runoff water. It typically occurs evenly over a uniform slope and goes unnoticed until most of the productive topsoil has been lost. Deposition of the eroded soil occurs at the bottom of the slope or in low areas.

Lighter-coloured soils on knolls, changes in soil horizon thickness and low crop yields on shoulder slopes and knolls are other indicators.

Rill Erosion

Rill erosion is the removal of soil by concentrated water running through little streamlets, or headcuts. Detachment in a rill occurs if the sediment in the flow is below the amount the load can transport and if the flow exceeds the soil's resistance to detachment. As detachment continues or flow increases, rills will become wider and deeper. Rill erosion mainly occurs as a result of concentrated overland flow of water leading to the development of small well-defined channels. These channels act as sediment sources and transport passages, leading to soil loss.

Gully Erosion

Gully erosion is "the removal of soil or soft rock material by water, forming distinct narrow channels, larger than rills, which usually carry water only during and immediately after rains". Gully erosion is an advanced stage of rill erosion.

A gully is a distinct channel, carved into a hillslope or valley bottom by intermittent or ephemeral runoff. Such channels are carved where the force exerted by flowing water – a function of its mass and velocity – exceeds the subsoil's resistance. Gully erosion results in significant amounts of land being taken out of production and creates hazardous conditions for the operators of farm machinery.

Bank Erosion

Bank erosion is the wearing away of the banks of a stream or river. This is distinguished from erosion of the bed of the watercourse, which is referred to as scour. Natural streams and constructed drainage channels act as outlets for surface water runoff and subsurface drainage systems. Bank erosion is the progressive undercutting, scouring and slumping of these drainage ways.

There are three main processes that cause bank erosion (scour, mass failure and slumping), and it is essential to determine which are operating at any particular site because the management required to slow or prevent them may differ. Bank scour is the direct removal of bank materials by the physical action of flowing water and is often dominant in smaller streams and the upper reaches of larger streams and rivers. Mass failure, which includes bank collapse and slumping, is where large chunks of bank material become unstable and topple into the stream or river in single events. Mass failure is often dominant in the lower reaches of large streams and often occurs in association with scouring of the lower banks.

Effects of Water Erosion on Site

The main on-site impact is the reduction in soil quality which results from the loss of the nutrient rich upper layers of the soil, and the reduced water-holding capacity of many eroded soils. The breakdown of aggregates and the removal of smaller particles or entire layers of soil or organic matter can weaken the structure and even change the texture.

Textural changes can in turn affect the water-holding capacity of the soil, making it more susceptible to extreme conditions such as drought. Crop emergence, growth and yield are directly affected by the loss of natural nutrients and applied fertilizers. Seeds and plants can be disturbed or completely removed by the erosion. Organic matter from the soil, residues and any applied manure, is relatively lightweight and can be readily transported off the field, particularly during spring thaw conditions. Pesticides may also be carried off the site with the eroded soil. Soil quality, structure, stability and texture can be affected by the loss of soil.

Effects of Water Erosion Off-Site

In addition to its on-site effects, the soil that is detached by accelerated water or wind erosion may be transported considerable distances. This gives rise to 'off-site problems'. Water erosion's main off-site effect is the movement of sediment and agricultural pollutants into watercourses. This can lead to the silting-up of dams, disruption of the ecosystems of lakes, and contamination of drinking water. In some cases, increased downstream flooding may also occur due to the reduced capacity of eroded soil to absorb water. Sediment can accumulate on down-slope and contribute to road damage. Sediment that reaches streams or watercourses can accelerate bank erosion, obstruct stream and drainage channels, fill in reservoirs, damage fish habitat and degrade downstream water quality.

Pesticides and fertilizers, frequently transported along with the eroding soil, contaminate or pollute downstream water sources, wetlands and lakes. Rapid bank erosion leads to loss of valuable land, reduced water quality as sediment and nutrients enter the stream, as well as threatening infrastructure such as roads, bridges and buildings. Stream bank erosion is the dominant source of sediment in many river systems.

Wind-Borne soil Erosion

Wind erosion is the detachment and transportation of soil particles by wind when the airstream passing over a surface generates sufficient lift and drag to overcome the forces of gravity, friction and cohesion. Once a particle has been dislodged from the surface, it may be transported in suspension or by saltation or by surface creep. Loss of topsoil by wind erosion over a relatively short time period can significantly decrease soil fertility and crop yield. The rate and magnitude of soil erosion by wind is controlled by the following factors:

- Soil Erodibility
- Soil Surface Roughness
- Climate (wind patterns, precipitation, frost action)
- Unsheltered Distance
- Vegetative Cover
- Topography (exposure, elevation, terrain roughness, localised funnelling of wind)
- Cultural practices (cultivation, vegetation depletion)

Soil Erodibility

Very fine soil particles are carried high into the air by the wind and transported great distances (suspension). Fine-to-medium size soil particles are lifted a short distance into the air and drop back to the soil surface, damaging crops and dislodging more soil (saltation).

Larger-sized soil particles that are too large to be lifted off the ground are dislodged by the wind and roll along the soil surface (surface creep). The abrasion that results from windblown particles breaks down stable surface aggregates and further increases the soil erodibility.

Soil Surface Roughness

Soil surfaces that are not rough offer little resistance to the wind. However, ridges left from tillage can dry out more quickly in a wind event, resulting in more loose, dry soil available to blow. Over time, soil surfaces become filled in, and the roughness is broken down by abrasion. This results in a smoother surface susceptible to the wind.

Climate

The speed and duration of the wind have a direct relationship to the extent of soil erosion. Soil moisture levels are very low at the surface of excessively drained soils or during periods of drought, thus releasing the particles for transport by wind.

Unsheltered Distance

A lack of windbreaks (trees, shrubs, crop residue, etc.) allows the wind to put soil particles into motion for greater distances, thus increasing abrasion and soil erosion. Knolls and hilltops are usually exposed and suffer the most.

Vegetative Cover

The lack of permanent vegetative cover in certain locations results in extensive wind erosion.

Loose, dry, bare soil is the most susceptible; however, crops that produce low levels of residue (e.g., soybeans and many vegetable crops) may not provide enough resistance. In severe cases, even crops that produce a lot of residue may not protect the soil.

Effects of Wind Erosion

Wind erosion damages crops through sandblasting of young seedlings or transplants, burial of plants or seed, and exposure of seed. Crops are ruined, resulting in costly delays and making re-seeding necessary. Plants damaged by sandblasting are vulnerable to the entry of disease with a resulting decrease in yield, loss of quality and market value. Soil drifting is a fertility-depleting process that can lead to poor crop growth and yield reductions in areas of fields where wind erosion is a recurring problem.

Continual drifting of an area gradually causes a textural change in the soil. Loss of fine sand, silt, clay and organic particles from sandy soils serves to lower the moisture-holding capacity of the soil. Also, soil nutrients and surface-applied chemicals can be carried along with the soil particles, contributing to off-site impacts. In addition, blowing dust can affect human health and create public safety hazards.

Tillage Erosion

Tillage erosion is the redistribution of soil through the action of tillage and gravity. It results in the progressive down-slope movement of soil, causing severe soil loss on upper-slope positions and accumulation in lower-slope positions. This form of erosion is a major delivery mechanism for water erosion. Tillage action moves soil to convergent areas of a field where surface water runoff concentrates. Also, exposed subsoil is highly erodible to the forces of water and wind.

Tillage erosion has the greatest potential for the "on-site" movement of soil and in many cases can cause more erosion than water or wind. The rate and magnitude of soil erosion by tillage is controlled by the following factors:

- Type of Tillage Equipment
- Direction
- Speed and Depth
- Number of Passes

Type of Tillage Equipment

Tillage equipment that lifts and carries will tend to move more soil. As an example, a chisel plow leaves far more crop residue on the soil surface than the conventional mold board plow but it can move as much soil as the mold board plow and move it to a greater distance. Using implements that do not move very much soil will help minimize the effects of tillage erosion.

- Direction: Tillage implements like a plow or disc throw soil either up or down slope, depending on the direction of tillage. Typically, more soil is moved while tilling in the down-slope direction than while tilling in the up-slope direction.

- Speed and Depth: The speed and depth of tillage operations will influence the amount of soil moved. Deep tillage disturbs more soil, while increased speed moves soil further.
- Number of Passes: Reducing the number of passes of tillage equipment reduces the movement of soil. It also leaves more crop residue on the soil surface and reduces pulverization of the soil aggregates, both of which can help resist water and wind erosion.

Effects of Tillage Erosion

Tillage erosion impacts crop development and yield. Crop growth on shoulder slopes and knolls is slow and stunted due to poor soil structure and loss of organic matter and is more susceptible to stress under adverse conditions. Changes in soil structure and texture can increase the erodibility of the soil and expose the soil to further erosion by the forces of water and wind.

Conservation Measures

Soil conservation is the preventing of soil loss from erosion or reduced fertility caused by over usage, acidification, salinization or other chemical soil contamination. Soil conservation is about solving the problems of land degradation, particularly soil erosion. Soil conservation is fundamentally a matter of determining a correct form of land use and management.

Soil conservation can be defined as the combination of the appropriate land use and management practices that promotes the productive and sustainable use of soils and, in the process, minimizes soil erosion and other forms of land degradation. Slash-and-burn and other unsustainable methods of subsistence farming are practiced in some lesser developed areas. A sequel to the deforestation is typically large scale erosion, loss of soil nutrients and sometimes total desertification. Techniques for improved soil conservation include crop rotation, cover crops, conservation tillage and planted windbreaks and affect both erosion and fertility.

Notable Methods of Soil Erosion Control

- Contour ploughing
- Terracing or terrace farming
- Keyline design
- Perimeter runoff control
- Windbreaks
- Cover crops/crop rotation
- Soil-conservation farming
- Salinity management

References

- Sandy-soil-the-basics: copernicus-psd.com, Retrieved 14, July 2020
- Sandy-soils, management-of-some-problem-soils, soil-management: fao.org, Retrieved 25, March 2020

- Understanding-and-improving-clay-soil-2539857: thespruce.com, Retrieved 03, August 2020
- Weathering, mountain-building, geolearning: geo.fuberlin.de, Retrieved 19, April 2020
- Biological-weathering-5633317: sciencing.com, Retrieved 06, February 2020
- Fundamentals-of-soil-stabilization-321727541: researchgate.net, Retrieved 11, May 2020
- Soil-fauna-diversity-function-soil-degradation-biological-indices-soil-restoration, biodiversity-conservation-and-utilization-in-a-diverse-world: intechopen.com, Retrieved 22, July 2020
- Soil-Erosion-Causes-and-Effects-314500264: researchgate.net, Retrieved 13, January 2020

2

Soil Science and its Branches

The branch of science which is concerned with the study of soil and its various aspects including its classification, formation and its physical, chemical and biological properties is known as Soil science. It has various branches like soil physics, soil chemistry and soil biology. The topics elaborated in this chapter will help in gaining a better perspective about the branches of soil science.

Soil science is the study concerned with observing and describing, collecting, establishing and systematizing facts, principles and methods in order to acquire an in depth knowledge of the soils, their properties and potential for production and conservation. Soil science uses an integrated multidisciplinary approach in the sense that it borrows concepts, techniques and processes from other sciences, but with a focus on soils. Soil science relies on 7 major supporting sciences or sub-disciplines:

- Climatology which affects the pedo-climate in terms of moisture and temperature conditions in the solum, and thus influences physical and chemical soil processes and plant and animal life.
- Geology which determines the nature and constitution (mineralogy) of the parent material from which the soil profile develops.
- Geomorphology (landform evolution) and hydrology which have a major impact on runoff, erosion and sedimentation processes, and differential warming up of soils.
- Physics, the basic laws of which determine the nature, intensity and interrelationships between the solid, liquid and gaseous soil components.
- Chemistry, concerned with the chemical constitution, chemical properties and chemical reactions in the soil, and their direct effect on soil fertility and nutrient supply to plants.
- Soil (micro) biology dealing with the soil fauna, the vegetation above and below the soil surface, as well as the microscopic soil population, and their role in various transformations and the liberation of nutrients.

Soil science as a discipline is relatively young compared to other sciences like mathematics or astronomy, the origin of which dates back for more than 2 000 years. The first who started to systematically study soils was the Russian geographer Dokouchaiev in the late 19th century, but his work became only known at the international floor after it had been published in German and disseminated by Marbut and Jenny in the USA.

The disastrous Dust Bowl in the Mid-West in the 1930s gave an additional impetus to the study of soils and was the start of the Soil Conservation Service in the US. Elsewhere in the world the growing interest in soils work received a major push after World War II when in a number of European countries a national soil survey institute was established while in the tropics the rapid development and increased demand of plantation crops asked for an better knowledge and understanding of soil-plant relations in these areas.

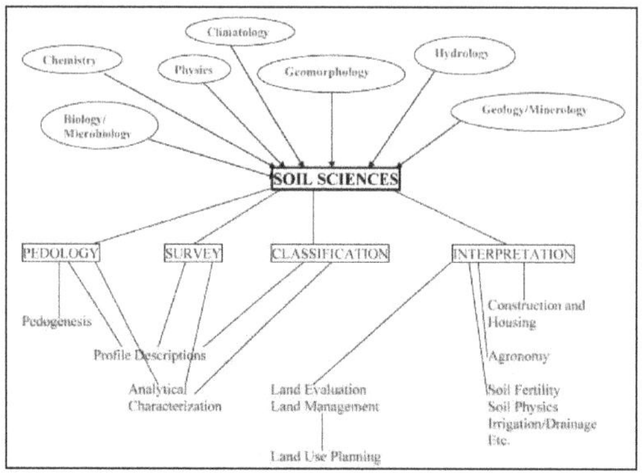

Supporting basic sciences and current sub-disciplines of soil science.

At present, there are four major disciplines within soil science focusing on different applications and users:

- Pedology, focusing on the formation (pedogenesis) and development of soils as recognized in the characteristics of the soil profile. It deals with fundamental and academic research aspects of soils, and includes the description and analytical characterization of the soil.
- Soil survey which describes the soil properties and delineates the geographical distribution of the different soils.
- Soil classification which organizes the soils and their particular properties on the basis of a hierarchical system of pre-defined criteria and classes. Though these criteria may vary as a function of the objectives, most international classification systems have a pedogenetic background.
- Applied soil sciences which interpret the soil properties in function of their ranking and potential for a specific objective. This includes applications (1) for housing and construction, focusing mainly on physical and much less on chemical properties; (2) for agronomic and crop production objectives, including soil fertility, suitability for drainage and irrigation, etc.; (3) for soil conservation and the protection of soils against physical loss by erosion or by chemical deterioration; and (4) land evaluation which assesses the production and use potential of the soil, including the development and monitoring of land use practices. The latter domain has since the 1980s gradually become a key issue for land resource management and land use planning.

Soil Biology

Soil biology is the study of soil biota and the interactions they have with each other and their environment. Soil biota includes four broad groups, based on size:

- Microflora (e.g. Bacteria, archaea, fungi and viruses <5 μm)
- Microfauna (e.g. Nematodes 10 μm-2 mm and protozoa 5 μm - 200 μm)
- Mesofauna (e.g. Small arthropods like mites and collembola 100 μm - 2 mm)

- Macrofauna (e.g. Earthworms and insects 2 mm - 2 m)

The soil biota comprises an enormous diversity with reports suggesting there could be greater than 15 000 different species per gram of soil. Much of this diversity, largely from the microflora group, is yet to be classified, however genomic technologies are helping us identify previously unknown soil organisms. Soil biology is represented by all three kingdoms of life: Bacteria, Archaea and Eukarya; in nearly all evolutionary branches. With such phylogenetic diversity it goes hand in hand that soil biology also possess an incredible range of functional diversity.

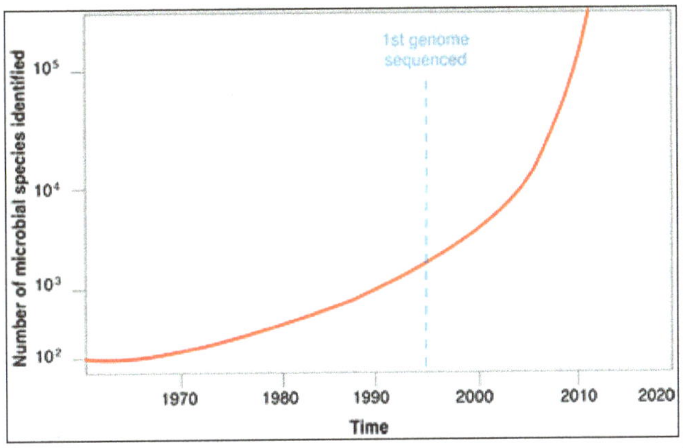

This diagram highlights how using genomic technologies based on DNA have exponentially increased the number of microbial organisms we have been able to identify.

There are three levels of participation by soil biota in natural soil processes/functions:

- Ecosystem engineers (e.g. earthworms, termites & ants)
- Litter transformers (e.g. microarthropods)
- Micro-food webs (e.g. microbes and microfaunal predators)

Ecosystem Engineers

Centipede

Ecosystem engineers such as ants and earthworms primarily alter the physical structure of soil but also have an influence on the overall rates of nutrient cycling and energy flows. These organisms initiate fragmentation of organic residues and take organic matter deeper into soil profiles. They also create pores that allow water and plant roots (as well as other soil biota) access to deeper parts of the profile.

Litter Transformers

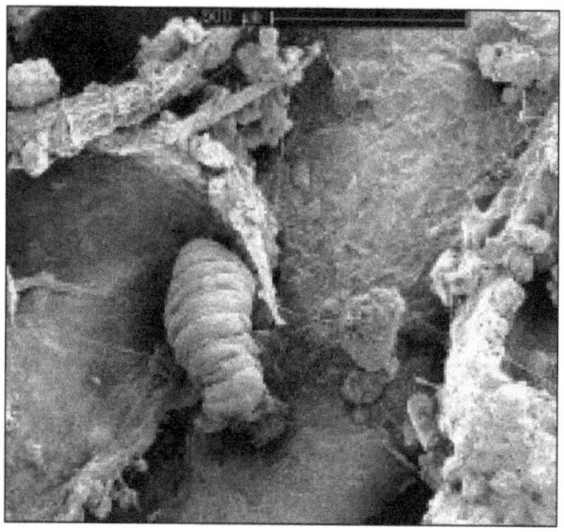

Collembola 'grazing' on decomposing plant material.

Litter transformers such as the microarthropod collembolas are involved in fragmenting plant residues and other organic substances making this material more available to microbes by increasing residue surface area for further chemical degradation and nutrient cycling.

Micro-Food Web Processors

Soil biota play a major role at all levels in micro-food webs whereby energy is transferred from one organism to another and as a consequence nutrients are recycled through the soil environment. For example, bacteria, archaea and fungi decompose plant litter through enzyme degradation to sequester nutrients and these organisms are in turn eaten by predatory protozoa, nematodes and arthropods. This predation helps maintain specific populations and keeps the predator prey balance in check. At the same time, however, litter transformers are breaking down dead organic matter, including remains of biota from all levels of the food web.

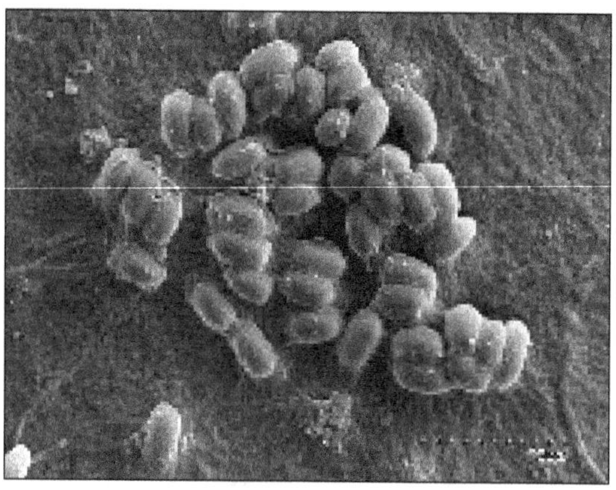

Bacterial colony.

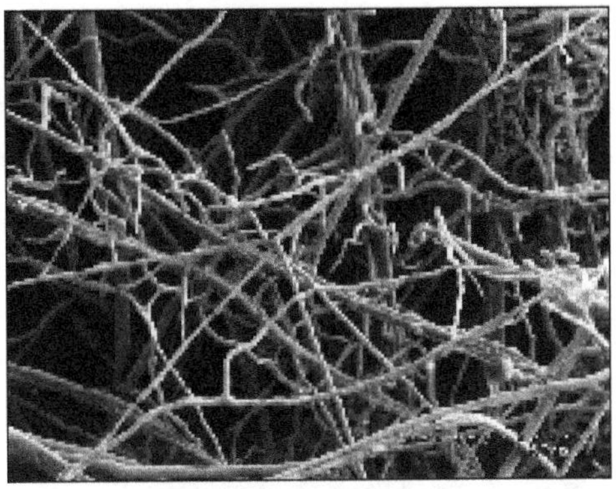

Fungal hyphal network.

Fungal spores forming at the terminus of hyphae.

Soil biology is not evenly distributed in soil and occurs in 'hot-spots' associated with soil organic matter. Decomposing residues, (obvious sites for both physical and enzymatic attack by a range of soil biota), the rhizosphere, (the soil zone that surrounds and is influenced by the roots of plants) and macroaggregates are all examples of such hot spots.

Fungal hyphae enmeshing decomposing straw residue where the fungi will secrete a suite of enzymes to break down the structure of the straw.

Importance of Soil Biology

There are three components of soil biology for land managers to consider; i) the natural processes themselves (soil engineering, litter transformation and micro-food web participation), ii) the impact of management on these processes, and iii) the combined effect of the first two on plant performance). The processes that are mediated by soil biology impact on plant (and animal) production systems by modifying the soil physical, chemical and biological environment within which plants grow and persist.

Decomposing Plant Residues

The whole process is very dynamic – while some of the residues are being eaten and fragmented for the first time by the litter transformers, other residues have already been sequestered by soil microflora, to be in turn eaten by microfauna predators.

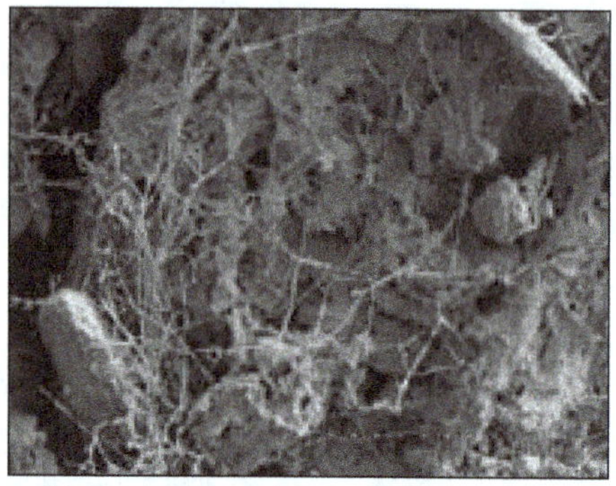

Fungal hyphae enmeshing organic residue which has been degraded by a suite of enzymes releasing nutrients into the soil matrix.

The process of converting plant residues (and other organic residues/wastes) into humus is facilitated by enzymes, either contained within the soil organisms or secreted by either living or dead soil organisms into the soil matrix. Some of the most common enzymes found in soils include cellulase (converts cellulose to glucose subunits), protease (converts protein to amino acids), lignase (breaks down lignin chains), urease (converts urea to ammonia and carbon dioxide) and glucosidase (converts starch to glucose). Eventually, the residues will have been processed to the point that they are relatively stable as humic substances (humic and fulvic acids). Humus represents the recalcitrant compounds of the original residues as well as the faeces and dead bodies of the soil organisms.

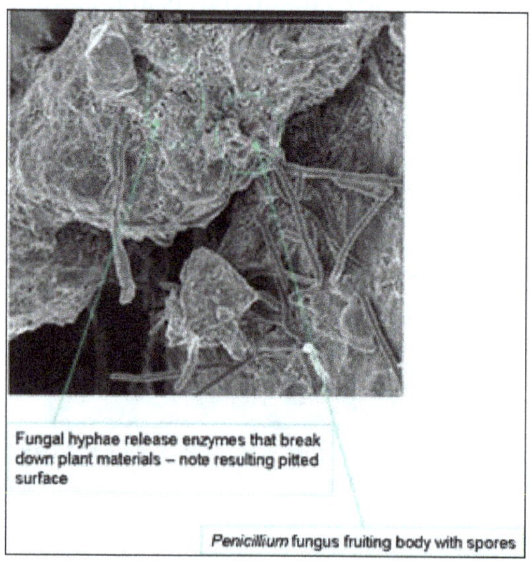

Fungal hyphae grow on plant residues - secreting a range of enzymes (such as cellulases, lignases and proteases) that break down these residues and release plant nutrients and gases (note pitted holes in substrate). Bacterial colonies remain attached to plant and soil surfaces and also secrete enzymes to release plant nutrients. The nutrients produced can be re-used by the microbes and are then respired or transformed into humus, taken up by living plants or lost through volatilisation or leaching.

Carbon Cycle

Soil biota play a key role in the carbon cycle whereby these organisms assimilate carbon (mainly from plant residues) and act as a large carbon sink as well as providing a substrate carbon source upon their death for other soil biota higher in the food web. The ability of soil microbes to tie up large quantities of carbon in these 'sinks' is becoming increasingly important in a world where greenhouse gas emissions are at the forefront of attitudes and policies alike.

Regulates Nutrient Supply

Nitrogen (N) Cycle

Soil biology mediates many different transformations of nitrogen in the entire N cycle. N-mineralisation and nitrification provide a useable inorganic N source for plant uptake while N fixing bacteria convert gaseous nitrogen (N_2) to a useable form. These processes are fundamental to the cycle and are of obvious importance in an agroecosystem context. The ability of soil denitrifiers to release nitrogen oxides from the soil also makes these microbes important in the context of greenhouse gas emissions. Nitrification is the biological oxidation (of ammonia) with oxygen into nitrite followed by the oxidation of these nitrites into nitrates that can be used by plants. The oxidation of ammonia into nitrite is performed by two groups of organisms, ammonia oxidizing bacteria and ammonia oxidizing archaea. Ammonia oxidizing bacteria can be found among the β- and α-proteobacteria. In soils the most studied ammonia oxidizing bacteria belong to the genera Nitrosomonas. The second step (oxidation of nitrite into nitrate) is mainly carried out by bacteria of the genus Nitrobacter.

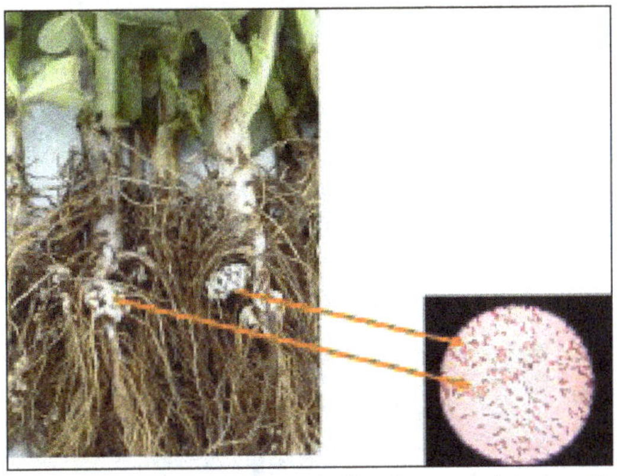

The "pink" nodules on legume roots where N-fixing bacteria live in a symbiotic relationship with the host plant. The "pink" colour indicates the bacteria are actively fixing N. "White" coloured nodules might indicate the bacteria are not fixing any N.

Nitrogen Fixation

Nitrogen fixation is the conversion of atmospheric nitrogen (N_2) into ammonia (NH_3). A number of soil bacteria contain the enzyme nitrogenase that allows them to fix N_2, either when free-living (Azospirillum and Azobacter) in the soil or in association with leguminous plants (Rhizobia and Bradyrhizobia). In agriculture, the association of Rhizobia and crop and pasture legumes plays a key role in land productivity. The Rhizobia colonise the legume roots and live in nodules that are formed on the roots. The Rhizobia are provided with habitat and supplied with nutrients and energy in the form of carbon compounds by the legume. In return, they fix atmospheric N_2 into ammonia, which is then converted to amino acids and amides within the nodules before being transported in the xylem sap to other plant parts. The amino acids and amides are then used by the legume for growth. This mutually beneficial arrangement is called symbiosis. The fixed N used for legume growth results in reducing the need to apply N fertiliser, and the N-rich residues that remain after the legume has been harvested or grazed add N to the soil. The actual amount of N fixed is determined by how well the legume crop or pasture grows, reflecting the effectiveness of nodulation, seasonal conditions and management, and the level of nitrate in the soil at planting. Soil nitrate suppresses nodulation and N_2 fixation. Thus, high soil nitrate means low N_2 fixation. Subsequently, in an effort to utilise this potential, screening for 'elite' Rhizobia is occurring in order to isolate effective N-fixing strains for newly released legume cultivars (including natives) as well as efficacious N-fixing strains for stressed soil conditions & dry-sowing (drought tolerant). Research is also being undertaken to optimise 'delivery' systems where carrier materials and preparations such as peat, granules and polymers as well as furrow injection technologies are being investigated.

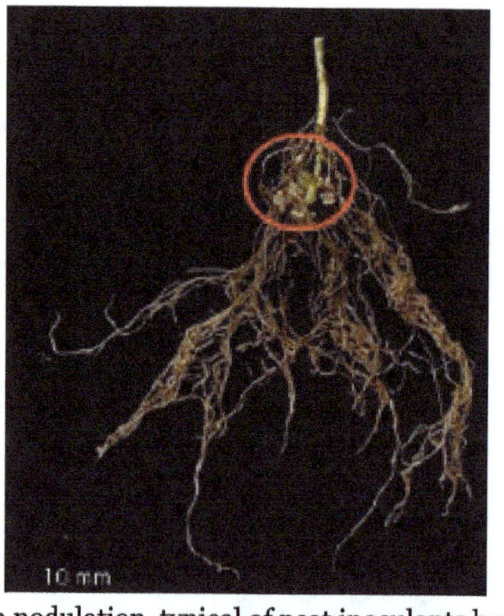

Crown nodulation, typical of peat inoculants but less efficient at N-fixation.

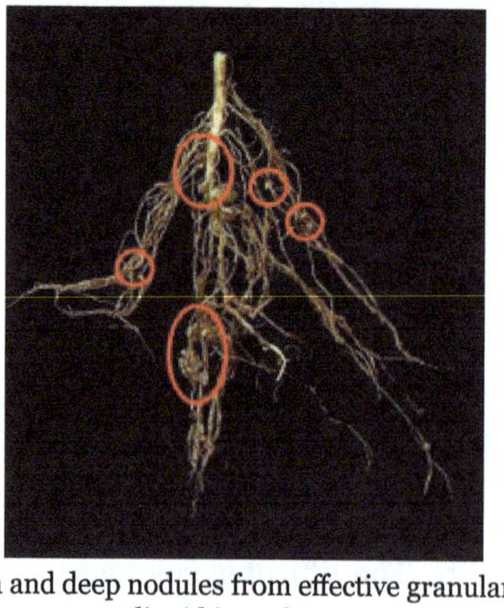

Crown and deep nodules from effective granular and liquid inoculants.

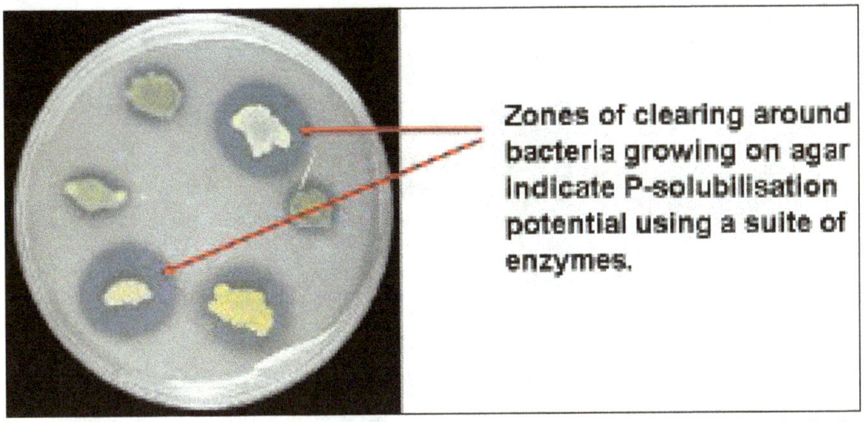

This figure highlights the enzymatic potential of many soil microbes to access "locked-up" P in the soil matrix therefore making the P available for plants to utilise.

Phosphorus Supply

Soil biology is crucial in the functioning of the phosphorus cycle. Different organisms are responsible for the mineralisation of organic phosphorus from organic to "useable" inorganic forms suitable for plant uptake. 70 - 80% of identifiable free living soil microbes, particularly bacteria and fungi, produce enzymes involved in phosphorus mineralisation (e.g. phosphatase, phytase, etc). Large quantities of phosphorus are also immobilised by soil microbes as it is a major component of two of the most important molecules in all living things: ATP (natures 'fuel') and DNA (the building block of life).

Mycorrhizas

Mycorrhizas are the associations plant roots form with fungi and are more often than not beneficial for both partners. The most common type of mycorrhizas is the arbuscular mycorrhizas (AM) involving members of the phylum Glomeromycota. Many plant species from both natural and agricultural ecosystems form these mutually beneficial symbiotic associations.

The AM fungi form a hyphal link between the host root and the soil. The fungi receive carbon sources as energy from the host plant and in return the extensive fungal hyphal network assimilates and transports nutrients particularly phosphorus and to a lesser extent micronutrients such as copper and zinc back into the plant roots from the soil. Arbuscules (which give the mycorrhizas their name) are "tree-like" branched hyphal structures formed in the host root cells and are considered the major site where carbon and nutrient exchange between plant and fungus occurs. AM increase the volume of soil explored by the plant well beyond the root zone, a trait that is especially relevant in Australian soils which are characteristically low in phosphorous. It is widely accepted that AM also increases water use potential by the host plant, improve the hosts ability to deal with salinity and help suppress some common pathogens.

Conventional agriculture practices, such as tillage, heavy fertilizer and fungicide use and poor crop rotations hinder the ability of plants to form symbiotic associations with AM fungi. There is great potential for low input agriculture to manage the system in a way that promotes mycorrhizal symbiosis, although all agroecosystems can benefit by promoting AM establishment. Proper management of AMF in the agroecosystems can improve the quality of the soil and the productivity of the land. Agricultural practices such as reduced tillage, low phosphorus fertiliser usage and rotations promote beneficial mycorrhizal symbiosis.

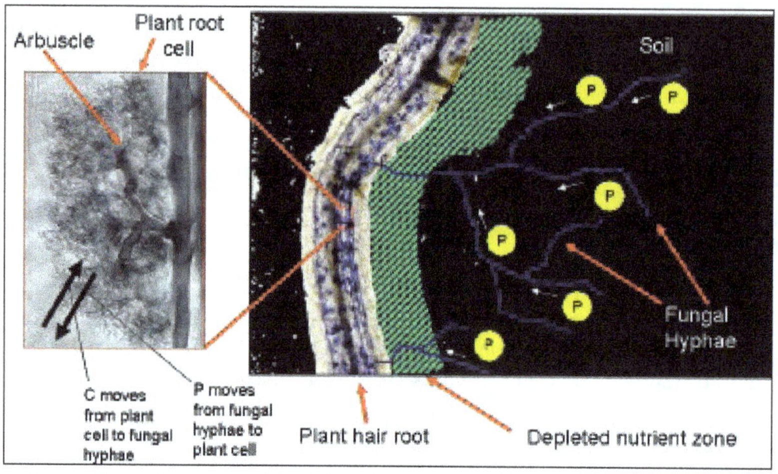

This figure shows how arbuscular mycorrhizal fungi form a symbiotic association with many crop species. The fungi increase the plant hosts access to P and in return obtain a C source from the plant.

Improved Soil Structure

Soil biota can have a direct effect on soil structure. Soil aggregates are clusters of soil particles comprising sand, silt, clay particles, humus and soil organisms. Aggregation of the soil is important for maintaining porosity and for stability of surface soils against erosion. The stability of aggregates is related to the level of soil organic matter and to the activity of soil organisms. For example, filamentous fungi, particularly mycorrhizal fungi, bind particles of soil into stable aggregates through physical entanglement. Bacteria also help stabilise smaller aggregates (2-250 μm) due to attractive forces of chemical charges on their surface. These smaller aggregates may subsequently be bound together into larger aggregates (>250 μm) through the action of fungal hyphae. Soil biota bi-products can also have a direct effect on soil structure. For example, glomalin is a glyco-protein produced by the common root-dwelling AM fungi, Glomerales, binds soil particles and organic matter to form soil aggregates – this compound is referred to as the "soils superglue". Polysaccharide exudates from bacteria also act as "super-glue" holding soil particles together to form aggregates.

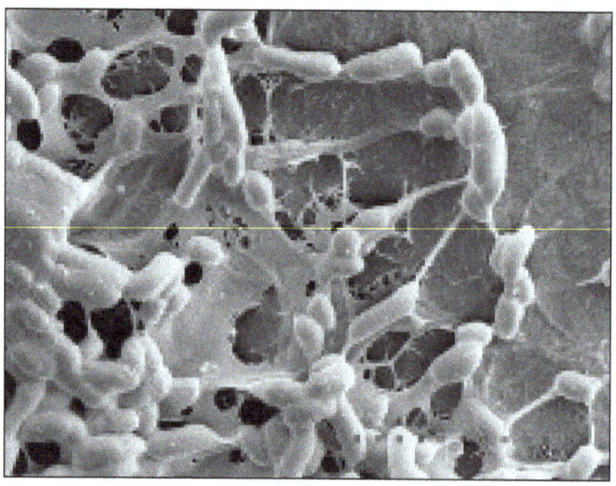

Bacterial "glue" composed of polysaccharide exudates.

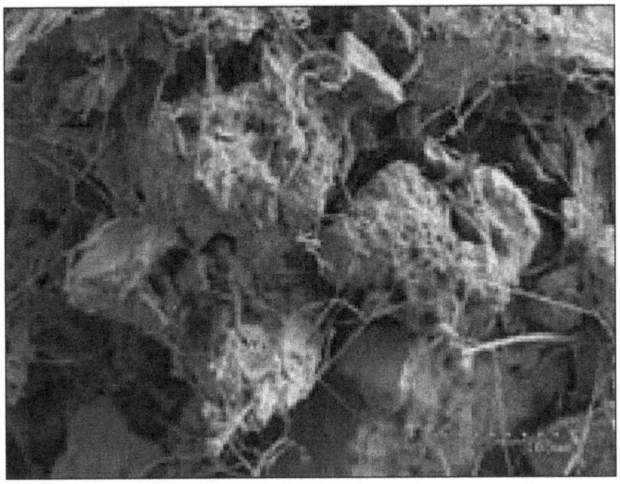

Fungal hyphae enmeshing soil aggregates. This helps form pores (macro and micro) that increase aeration and water infiltration in soil.

Degrades Chemicals Inputs

Agricultural pesticides (herbicides, insecticides and fungicides) are degraded in the soil principally by the action of microorganisms, a process termed biodegradation, where biodegradation is defined as the breakdown of a substance to smaller products caused by microorganisms or their enzymes. The rate at which different pesticides are biodegraded varies widely but generally degradation is faster in soils with high microbial diversity and activity. Some pesticides such as DDT have proven to be recalcitrant. Consequently they remain in the environment for a very long time and are known to accumulate into food chains decades after their application to soil. Pesticides which are more readily biodegradable, such as the organophosphates, are now used in preference to the more persistent chlorinated pesticides. Others, such as atrazine and simazine, are biodegradable at slow rates and may be leached from soil to ground water, where they pose a threat to drinking water supplies. At the other end of the scale, there are pesticides, such as carbofuran and diazinon that are readily biodegraded, and are broken down so rapidly under certain soil conditions that they may not allow effective pest control. Furthermore, toxic elements, e.g. arsenic, chromium and mercury, can be 'locked' in the soil by microbial activity, preventing further pollution.

Biocontrol

Many soils support microbes that are deleterious to plant growth; the symptoms can be visible (clinical) or invisible (sub-clinical). Soil fumigation has quantified the link between soil microbes with crop yield performance. A variable impact on grain yields has been demonstrated with up to an 80% increase in yield on highly alkaline Calcarosols (D. Roget) & up to 60% reduction in yields on highly acidic Dermosols (P. Mele). Why the variability in response? – fumigation could actually knock out beneficial organisms in a disease suppressive soil allowing a disease causing agent to proliferate in the soil. Organisms identified as potential biocontrol agents in suppressive soils include the soil fungi, Trichoderma spp, Penicillium spp and Metarhizium spp and the bacteria, Pantoea, Exiguobacterium and Microbacterium. Reported modes of action of the suppressive microorganisms are production of antibiotics, competition for infection sites on the roots, competition for space and nutrients in the rhizosphere and parasitism of the pathogens. These beneficial organisms should be encouraged through management such as crop rotations and maximizing the return of residues (carbon source) to the soil in order to restore a balance between the soil biota in the soil so that pathogens do not dominate. Disease suppressive soils are more frequently being identified and soil-borne pathogens such as Rhizoctonia, GG Take-all & Pythium can be controlled in these soils. A suppressive soil is one that, despite having conditions that favour the establishment of a disease-forming organism, results in the pathogen:

- Either not establishing.
- Establishing, but not producing disease symptoms.
- Establishing and then declining.

A suppressive soil does not result in the elimination, but the control of disease forming organisms. Farmers should be trying to achieve a level of suppression (microbial balancing) that prevents disease. But suppression is relative - a soil may be suppressive, but disease at a lower level is still experienced.

Greenhouse Gases

Primary industries are significant emitters of greenhouse gases. Agriculture is the third largest contributor to Victoria's total greenhouse gas emissions. Therefore, this sector has an important role to play in reducing Victoria's emissions.

Nitrous oxide emissions are produced by a range of identifying bacteria in the soil, where they convert nitrate into nitrous oxide. These losses are greatest when soils are warm and waterlogged, and in those with high nitrate contents. Thus it is important to apply nitrogen fertilisers only at times, and in quantities and forms, useful to plants as overuse of fertiliser can drastically increase levels of denitrification. Nitrous oxide is a more potent greenhouse gas than carbon dioxide. One unit of nitrous oxide is equivalent to 310 units of CO_2. Conventional tillage systems also release more CO_2 into the atmosphere than no-till systems as the incorporated residues turnover more quickly and result in more carbon being respired by the microbial community. No–till systems tend to lock up more carbon in the form of organic matter. A large soil microbial community can also tie up carbon and nitrogen that might otherwise be released into the atmosphere as greenhouse gases.

Soil Chemistry

Soil chemistry is the discipline embracing all chemical and mineralogical compounds and reactions occurring in soils and soil-forming processes. The goals of soil chemistry are: (1) to establish, through chemical analysis, compositional limits of natural soil types and optimal growth conditions for the various plant communities, (2) to derive principles governing development of soil profiles, and (3) to apply analytical and theoretical knowledge to the improvement of agricultural production.

Chemical Properties of Soil

The chemical side of a soil is extremely important of course and is about the correct balance of the available nutrients in the soil. This is largely determined by the organic-matter content and its humus percentage; this is the 'pantry' of nutrients on any farm. The extent to which minerals have a dominant presence or not, affect the release of specific nutrients. Supplementing shortages is important, but the right balance is even more important. The soil only produces nutrients if you have the right balance.

Agricultural chemistry committed an error by designating the most common elements as the most important. Fertilisers are a result of SSP thinking which is focused on just three elements, when a crop requires more than 40 elements to be able to flourish.

Balanced Soil Chemistry

It starts with the minerals balance in the soil. The task of the fertilisers is to balance the chemistry of the soil and to feed the crop. The ideal content of the clay-humus complex (CEC) is 60-70% calcium, 10-20% magnesium and 2-5% potassium, which always leaves 10% that can be filled with hydrogen. Hydrogen determines the pH of the soil and is crucial to keeping soil life active and creating availability of the nutrients.

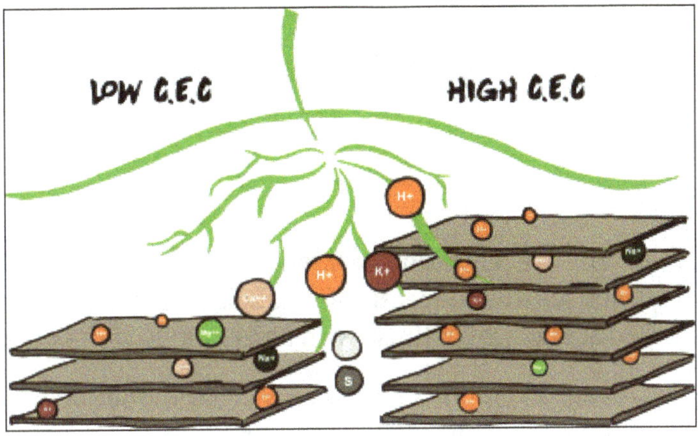

The chemical soil fertility is largely determined by the presence of macro-elements, the presence of micro-elements or trace elements, the acidity or pH, the salt content or EC and the cation-exchange capacity or CEC.

Macro-Elements

Macro-elements are the most important nutrients for plants. They occur naturally in the soil to some extent and can be supplemented with fertilisers, manure and compost. Legumes are able to bind nitrogen. The macro-elements are calcium (Ca), magnesium (Mg), potassium (K), sulphur (S), phosphate (P), nitrogen (N), sodium (Na).

Micro-Elements

Trace elements are nutrients that plants need in small doses. The trace elements are boron (B), copper (Cu), manganese (Mn), cobalt (Co), silicon (Si), zinc (Zn), iron (Fe) and molybdenum (Mo). Signs of shortage occur when the disappearance of trace elements through the crops is not compensated adequately with supplements by means of fertiliser, manure or compost, or when the availability of certain elements is limited by the pH or mineral imbalance in the soil. It is difficult to solve a lack of trace elements in the plant at source, as many shortages are the result of a shortage or surplus of another mineral in the soil. Crop analyses can help to trace an acute shortage and to solve it at plant level. Soil analyses provide insight into the total mineral balance which tells you where shortages can be expected, which can then be solved.

Acidity (pH)

Acidity (pH) is a measure of the concentration of free hydrogen ions (H^+). A high concentration in the soil signifies a low pH, whilst a low concentration equals a high pH. Soils and other substances with a pH below 6 are called acidic, whilst those with a pH above 6 are known as base or alkaline.

The form in which macro-elements and micro-elements occur in the soil depends on the pH. This means that the pH exercises a significant influence on the availability of nutrients for plants, as plants cannot absorb macro-elements and micro-elements in all their forms. A pH that is less than 4.5 restricts the availability of a number of elements. Soil-life activity and consequently the mineralisation of organic matter are limited in that situation.

Agricultural soils are often limed in order to keep the pH sufficiently high. Adding lime also stimulates soil life and promotes the breakdown or organic matter.

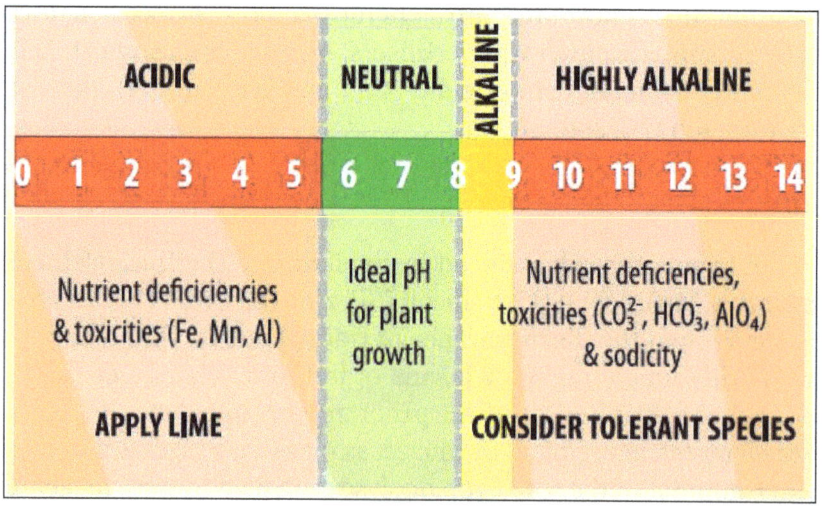

Salt Levels

The salt level is the sum of all the mineral salts that are present in the soil. They can originate from the soil itself, fertiliser, organic manure and in coastal areas from salt marshes or tidal marshes. When the salt levels in the soil are higher than in the cells of the plant roots, the moisture is drawn from the roots and the fine hair roots die off. Over time this impedes the moisture and nutrients absorption by the plant and causes reduced growth or death of the plant. Crops vary in the extent to which they tolerate salt and the salt levels at which they can still provide a good yield. Some crops grow well in salty soils.

CEC

CEC is the abbreviation of cation exchange capacity, which notes the amount of cations can be bound to the soil. These cations are mainly Ca^{2+}, Mg^{2+}, K^{+}, Na^{+} and to a lesser extent NH^{4+} , Al^{3+}, Fe^{2+}, Mn^{2+} and H^{+}. Plenty of calcium (Ca^{2+}) in the adsorption complex is important to the soil structure. Little calcium or lots of potassium or magnesium bound to clay provides a less good structure. Calcium keeps the clay platelets at a sufficient distance from each other and in doing so produces an airy structure. If there is a lot of potassium or magnesium between the clay platelets, the clay platelets come too close together and the soil is harder to work.

The balance between Ca and Mg is therefore crucial to the right structure of a soil, as it determines the amount of room there is for water and oxygen and that in turns determines the extent of the soil biology.

Soil Physics

Soil physics is the branch of soil science dealing with the physical properties of the soil, as well as with the measurement, prediction, and control of the physical processes taking place in and through the soil. Soil physics involves both the understanding of soil behavior and its management and is, therefore, both a basic and an applied science.

The soil is a mixture of mineral particles, organic substances, soluble components, water and gases composing the bio-substrate. The mineral fraction consists of particles of different size, shape and composition. Organic matter includes plant and animal residues in various stages of decomposition, but also living organisms. The soluble components determine the chemical composition of the soil water filling part or all of the pores between the solid particles. Pores that do not contain water are filled with gases or vapor. The nature and properties of the solid, liquid and gaseous phases affect the chemical and physical interactions, and these are influenced themselves by temperature, pressure and light.

The mineral fraction of the solid phase may be completely dispersed or stick together to granules and/or aggregates. The total amount of individual particles is referred to as soil texture while the aggregation of individual particles to soil clods (or soil peds) is known as soil structure.

The soil as a whole is a dynamic entity and is constantly in search of an equilibrium state, as it alternately wets and dries, swells and shrinks, disperses and flocculates, hardens and softens, warms and cools, freezes and thaws, compacts and cracks, absorbs and emits gases, adsorbs and releases exchangeable ions, precipitates and dissolves salts, becomes acidic or alkaline, and exhibits aerobic or anaerobic conditions leading to chemical oxidation or reduction. The soil is a complex system that is idealized to understand and simulate some aspects of the soil behavior.

Soil interacts also with the ambient bodies and thus, absorbs energy and water from the atmosphere. The large soil pores drain the water downward while the fine pores retain moisture either for plant consumption or evaporation. In this respect, the soil serves as a medium for plant growth. Plant roots can penetrate and ramify in a loose, porous and well-aerated soil to achieve anchorage and to extract moisture and nutrients.

Soil Texture

Soil texture is an intrinsic attribute of the soil and it is, therefore, often used to specify its physical composition. Soil texture refers to the size range of the solid particles in the soil resulting from the weathering of parent rocks. The loose products may stay in place to form a residual soil, or may be transported by water, glacier or wind and be deposited at some distance from the original site.

The process of soil formation requires many thousands of years. The size and shape of the individual particles and their chemical and mineralogical composition are of particular interest because they affect the nature and properties of the soils involved, as for example hydraulic conductivity and natural fertility.

Soil particles are grouped into classes on the basis of size, and the relative proportions of these classes characterize the texture of the soil. These classes often differ also in mineral composition. Both particle size and mineral composition largely determine the nature and behavior of a soil like porosity, interactions with solutes, soil compaction, soil strength etc.

The soil particle sizes are expressed in terms of an equivalent diameter because soil particles are not spherical but an-isometric. The equivalent diameter depends on the size of the particles. For larger soil particles that are separated by mechanical sieving it refers to the diameter of a sphere that passes through a given square opening size. For smaller particles sedimentation techniques are used to determine the sizes, and the equivalent diameter refers to the diameter of a sphere with the same density and velocity of settling in an aqueous medium. The largest soil particles are generally visible to the naked eye, whereas the smallest colloidal particles can only be observed with an electron microscope.

Soil Density

It is defined as the property of soil, suggesting how closely the soil grains will be arranged within the volume of soil. There can be spaces between particles, containing air and water. The typical density of sand is 2.6 Mg/m3. The two types of density are given as follows, 1) bulk density and 2) particle density.

Bulk density is considered as the total mass of the soil (includes water present) divided by soil volume. Particle density is the average density of soil particles, explained as the ratio of mass of soil solids to the volume. It refers to only the solids excluding the voids.

Factors Affecting Density of Soil

The density of the soil is dependent on soil texture. The soils like silty and clayey loams have lower bulk density values compared to that of the sandy soils. Presence of organic matter in the soil, aggregates, packing arrangements, and depth of the soil influence the density of the soil mass.

Significance

Density of soil reflects the ability of the soil to help towards the structural support, allowing the soil aeration and water movement through the pores. The highly dense soil will have decreased value of settlements. The values of density and unit weight of soils are used to increase shear strength of soil.

When the soil is compacted to increase the density, it will decrease permeability. High density fill can be used to improve sub-grade with low bearing capacity. Hence, it will thereby reduce time of construction.

Soil Color

Soil color has been found to be the property of soil that most reflects its pedogenic environment and history. Soil organic matter and iron oxides contribute most to soil color. Organic matter darkens soil, while iron oxides produce a range of soil colors that are dependent on the oxidation state of the iron.

Types of Soil Color

Clear or White Soil Color

They may be white from which they may be influenced by calcium and magnesium carbonates, gypsum or other more soluble salts. Often, a white layer, mostly quartz occurs between organic matter on the surface where pigments were removed (Wisconsin State Soil).

Brown Soil Color

Brown soils might be brown from decaying plant material. The darker color often indicates an increase in decomposed organic matter known as humus. Soil has living organisms and dead organic matter, which decomposes into black humus. In grassland (prairie) soils the dark color permeates through the surface layers bringing with it nutrients and high fertility Deeper in the soil, the organic pigment coats surfaces of soil, making them darker than the color inside. Humus color decreases with depth and iron pigments become more apparent. In forested areas, organic matter (leaves, needles, pine cones, dead animals) accumulates on top of the soil. Water-soluble carbon moves down through the soil and scavenges bits of humus and iron that accumulate below in black, humic bands over reddish iron bands.

Yellow or Red Soil Color

Yellow or red soil indicates the presence of oxidized ferric iron oxides. The red color might be mainly due to ferric oxides occurring as thin coatings on the soil particles while the iron oxide occurs as hematite or as hydrous ferric oxide, the color is red and when it occurs in the hydrate form as limonite the soil gets a yellow color.

Grey Soil Color

Organic matter plays an indirect, but crucial role in the removal of iron and manganese pigments in wet soils. All bacteria, including those that reduce iron and manganese, must have a food source. Therefore, anaerobic bacteria thrive in concentrations of organic matter, particularly in dead roots. Here, concentrations of grey mottles develop.

Soil Consistency

Soil Consistency the feel of the soil, reflecting relative resistance to pressure: e.g. friable, firm, hard, loose, plastic. The term soil consistency is used to describe the resistance of a soil at various moisture contents to mechanical stresses or manipulations. It is commonly measured by feeling and manipulating the soil by hand or by pulling a tillage instrument through it. The consistency of soils is generally described at three soil moisture levels: wet, moist and dry. Terms used to describe soil consistency at these three moisture levels are shown in the table.

Wet soils	Moist soils	Dry soils
Stickiness	Plasticity	Nonsticky
Nonplastic	Loose	Loose
Slightly sticky	Slightly plastic	Very friable
Soft	Sticky	Plastic
Friable	Slightly hard	Very sticky
Very plastic	Firm Hard	Very firm
Very hard	Extremely firm	Extremely hard

Terms such as weakly cemented, strongly cemented, and indurated are used to define categories of cementation. Consistency has importance for the practical use of soils such as soil tillage and compaction by farm machinery.

Acid basin clays are featured by heavy soil consistency (making the soils difficult to plough both when wet and when dry). Noncalcareous Grey Floodplain soils and Noncalcareous Dark Grey Floodplain soils show firmer consistency. Calcareous Brown Floodplain soils develop more friable consistency.

Soil Temperature

Soil temperature is the measure of how hot or cold the soil is. If you want to get more technical, it's the detection of the internal energy of the soil, or its heat. Soil scientists consider temperature a very important soil physical property and it controls many chemical and biological processes within the soil.

When a farmer plants, he waits for the temperature to rise above a certain value (i.e., he waits until the soil has enough internal energy) so that his seeds will germinate. For every 10 °C rise in soil temperature, we expect biological activities, like soil respiration, to double. If the temperature gets too high, it kills things that live in the soil, like plant roots and other organisms. Likewise, chemical reactions that aren't mediated by soil organisms tend to go faster as the temperature increases.

A gauge for measuring soil temperature.

Some interesting happenings in soil involving temperature are: 1) the flow of heat and 2) how soils "heat up—in other words, the amount of heat they must absorb to increase the soil temperature. In the first case, heat in the soil flows from places of high temperature to low temperature so heat can move into the deeper layers of the soil profile. Interestingly, because of the time it takes for heat to move, there is a depth in soils at which the highest temperature is in the winter and the lowest temperature is in the summer.

Soil Resistivity

The measure of the resistance offered by the soil in the flow of electricity is called the soil resistivity. The resistivity of the soil depends on the various factors like soil composition, moisture, temperature, etc. Generally, the soil is not homogenous, and their resistivity varies with the depth. The soil having a low resistivity is good for designing the grounding system. The resistivity of the soil is measured in ohmmeter or ohm-centimeters.

The resistivity of the soil mainly depends on its temperature. When the temperature of the soil is more than 0°, then its effect on soil resistivity is negligible. At 0° the water starts freezing and resistivity increases. The magnitude of the current also affects the resistivity of the soil. If the magnitude of current dissipated in the soil is high, it may cause significant drying of soil and increase its resistivity.

The resistivity of the soil varies with the depth. The lower layers of the soil have greater moisture content and lower resistivity. If the lower layer contains hard and rocky layers, then their resistivity may increase with the depth.

Measurement of Soil Resistivity

The resistivity of the soil is usually measured by the four spike methods. In this method the four spikes arranged in the straight line are driven into the soil at equal distance. A known current is passed between electrode C_1 and C_2 and potential drop V is measured across P_1 and P_2. The current I developed an electric field which is proportional to current density and soil resistivity. The voltage V is proportional to this field.

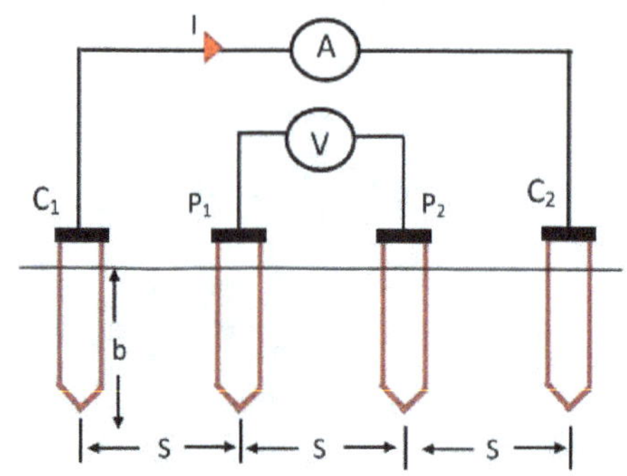

Measurement of Soil Resistivity

The soil resistivity is proportional to the ratio of the voltage V and current I and is given as:

$$\rho = \frac{\frac{4\pi SV}{I}}{1 + \frac{2S}{\sqrt{S^2 + 4b^2}} - \frac{2S}{\sqrt{4S^2 + 4b^2}}}$$

Where ρ is the resistivity of the soil and their unit is ohmmeters. S is the horizontal space between the spikes in m and b is the depth of burial in metre. If the measurement is to be carried out using the main supply, an isolating transformer should be connected between the main supply and the test setup. So that the result may not be affected by it.

References

- Soil-chemistry, science: britannica.com, Retrieved 17, March 2020
- Chemical-properties, soil-fertilization: healthy-soil.com, Retrieved 13, August 2020
- Density-of-soil: chegg.com, Retrieved 29, January 2020
- Soil-Consistency: banglapedia.org, Retrieved 17, May 2020
- Soil-resistivity: circuitglobe.com, Retrieved 25, January 2020

3

Soil Nutrition

Soil is rich in various nutrients which are essential for the growth of plants such as nitrogen, potassium and phosphorus. This chapter includes soil fertility, essential elements, macronutrients, micronutrients, sodium in soil, soil ion exchange soil pH, soil acidity, etc. All the diverse principles of soil nutrition have been carefully analyzed in this chapter.

Soil nutrient composition plays a key role in determining the goodness of a soil. A healthy soil will have all the essential elements in the right proportions to support healthy plant growth throughout its life cycle. However, when a soil has been cultivated it may require the addition of organic or inorganic fertilizers.

Soil Nutrient Sources

Sources of soil nutrient include:

- Organic matter decomposition
- Precipitation
- Biological nitrogen fixation
- Inorganic fertilizer application
- Weathering of soil rocks and minerals

What affects the availability of nutrients in soil? Factors that affect the availability of soil nutrients include leaching, soil erosion, soil pH, denitrification, volatilization, nitrogen immobilization and crop nutrient uptake.

Soil Erosion

Soil Erosion disrupts the soil structure, washes away organic matter in the soil and therefore reducing soil fertility. This often increases the need for additional and costly fertilizers to compensate for nutrient losses. Some causes of soil erosion include:

- Soil gradient: The steeper the gradient of a soil, the more vulnerable it is to soil erosion.
- Soil erodibility: The more erodible a soil, the more sensitive it is to erosion. This is influenced by the soil characteristic and nature. A soil that has experienced previous erosion will have higher chances of erosion.
- Vegetative cover: Soil vegetation can protect the soil from wind or water erosion by creating vegetative cover to the soil.

Leaching

This is the washing downward of nutrients in the soil below the root zone. Some of the factors influencing leaching include:

- Mobility of nutrients: When there is sufficient water in the soil, nutrients in soil solution can be easily washed down beyond the root zone. An example is nitrogen present in the form of nitrate; a highly mobile negatively charged ion.
- Soil texture: Leaching occurs in soils which have high water infiltration rates and low ability to hold nutrients. Examples of such are the sandy soil and clay soil.

Denitrification

This process results in the gaseous loss of nitrogen to the atmosphere. Denitrification occurs in warm and anaerobic (saturated) soils usually having high nitrate levels. During this process soil microbes break down nitrate to obtain oxygen for their respiration. The end product is nitrogen gas released to the atmosphere. Ways of avoiding loss of nitrogen through denitrification include proper timing of organic or inorganic fertilizer application; such that the soil receives it when it really needs it. Crop producers are advised to apply fertilizers in splits in order to match the crop demand for nitrogen with supply.

NO_3^- →	NO_2^- →	NO →	N_2O →	N_2
Nitrate	Nitrite	Nitric oxide	Nitrous oxide (gas)	Nitrogen gas

Stages of denitrification reaction.

Volatilization

This process also involves the gaseous loss of nitrogen to the atmosphere. When nitrogen fertilizers are applied in urea form (this could be inorganic fertilizers of animal manure), an enzyme urease catalyzes the reaction of urea with water resulting in ammonia gas released to the atmosphere. Volatilization occurs in warm and moist soil conditions. Volatilization also increases with high soil pH. To reduce nitrogen loss through this means, crop producers are encouraged to adapt methods of applying urea or ammonium based fertilizers below the soil surface.

Soil Fertility

Soil that produces abundant crops under the appropriate environmental conditions is coined the term "fertile". Soil fertility reflects the soil- plant relationship with reference to plant nutrients available in soil. Understanding the principles of soil fertility is vital to efficient nutrient management, profitable crop protection and environmental protection.

By definition, soil fertility is defined as the innate capacity of soil to provide adequate amount of nutrients essential for plant growth during favorable conditions such as temperature, water and light, and the physical condition of soil while soil productivity is the economic concept that signifies the capability of soil to produce specified plants under well-defined and specified systems of management inputs and environmental conditions. It is a function of several factors rather than being a soil property alone.

Types of Soil Fertility

Inherent or Natural Fertility

- The soil, as a nature contains some nutrients, which is known as inherent fertility.
- Among plant nutrients nitrogen, phosphorus and potassium is essential for the normal growth and yield of crop. The inherent fertility has a limiting factor from which the fertility is not decreased.

Acquired Fertility

- The fertility developed by application of manures and fertilizers, tillage, irrigation, etc., is known as acquired fertility.
- The acquired fertility has also a limiting factor. It is found by experiment that the yield does not increase remarkably by application of additional quantity of fertilizers.

Factors Effecting Soil Fertility

The factors that are affecting soil fertility may be of two types:

- Natural Factors: The natural factors are those which influence the soil formation and the artificial factors are related to the proper use of land.
- Artificial Factors: The factors effecting the fertility of soil are parent material, climate and vegetation, topography, inherent capacity of soil to supply nutrient, physical condition of soil, soil age, micro-organisms, availability of plant nutrients, soil composition, organic matter, soil erosion, cropping system and favorable environment for root growth.

Maintenance of Soil Fertility

Maintenance of soil fertility is a great problem of our farmers. Cultivation of particular crop year after year in the same field decreases the soil fertility. To increase the soil fertility, it is necessary to check the loss of nutrient and to increase the nutrient content of soil. The following things must be properly followed for increasing the fertility of soil:

- Proper use of land.
- Good tillage.
- Crop rotation.
- Control of weeds.
- Maintenance of optimum moisture in the soil.
- Control of soil erosion.
- Cultivation of green manure crops.

- Application of manures.
- Cultivation of cover crops.
- Removal of excess water (drainage).
- Application of fertilizers.
- Maintenance of proper soil reaction.

Essential Nutrients

For an element to be called as essential, it needs to meet the following criteria:

- The plant cannot complete its life cycle (seed to new seed) without it.
- If the element is specific in its physiological function in plants.
- If the malady that develops in plant can be remedied only by that element.
- The element's function cannot be replaced by another element.
- The element is directly involved in the plant's growth and reproduction.

Overall, 18 elements have been identified as essential for the growth of all plants. They are further classified not on their relative importance but on basis of their relative amounts into: Macronutrients (C, H, O, N, P, K, Ca, Mg, S) and Micronutrients (B, Cu, Fe, Mn, Zn, Mo, Cl, Co, Ni).

Macronutrients

The elements occurring in ample levels in plant biomass and fluids are known as macronutrients. They are divided into three classes: Structural (C, H, O), Primary (N, P, K) and Secondary (Ca, Mg, S).

Table: Macro-elemental nutrient, their uptake form and their plant mobility.

	Nutrient	Uptake form	Mobility in Plant	Mobility in Soil
1.	Carbon	CO_2, H_2CO_3		
2.	Hydrogen	H^+, OH^-, H_2O		
3.	Oxygen	O_2		
4.	Nitrogen	NO_3^-, NH_4	Mobile	Mobile as NO_3^-, immobile as NH_4^+
5.	Phosphorous	HPO_4^{2-}, $H_2PO_4^-$	Somewhat mobile	Immobile
6.	Potassium	K^+	Very mobile	Somewhat mobile
7.	Calcium	Ca^{2+}	Immobile	Somewhat mobile
8.	Magnesium	Mg^{2+}	Somewhat mobile	Immobile
9.	Sulphur	SO_4^-	Mobile	Mobile

Primary Macronutrients

Macronutrients are essential for plant growth and a good overall state of the plant. The primary macronutrients are Nitrogen (N), Phosphorus (P), and Potassium (K).

Nitrogen

Nitrogen is essential for plant development, since it plays a fundamental role in energy metabolism and protein synthesis. Nitrogen is absorbed by the plant in the form of a nitrate. This macronutrient is directly related to plant growth. It is indispensable for photosynthesis activity and chlorophyll formation. Nitrogen is involved, above all, in the aerial zone, the part of the plant that a person sees. It promotes cellular multiplication. A nitrogen deficiency results in a loss of vigor and color. Growth becomes slow and leaves fall off, starting at the bottom of the plant.

Phosphorus

Phosphorus is involved in root growth, which it stimulates. In the aerial zone it favors flowering. Although phosphorus is also necessary during the plant's growth period, it is much more involved in the flowering stage. Phosphorus is involved in transporting and storing energy. It improves the plant's general state and increases the plant's ability to withstand adverse climatological conditions. Phosphorus is essential in organic compound formation and the correct execution of photosynthesis. A shortage of phosphorous results in a late, deficient flowering, browning and wrinkling of the leaves, and a lack of vigor in general.

Potassium

Potassium is involved in the regulation of water and the transport of the plant's reserve substances. It increases photosynthesis capacity, strengthens cell tissue, and activates the absorption of nitrates. Potassium stimulates flowering and the synthesis of carbohydrates and enzymes. This, in turn, provides an increase in the plant's ability to withstand unfavorable environments such as low temperatures, and prevents withering. Therefore, a lack of potassium reduces plant resilience to dry spells and frosts or to a fungus attack. This, in turn, results in a lack of balance among other nutrients, such as calcium, magnesium, and nitrogen. When there is a potassium insufficiency, dark spots appear on the leaves.

Secondary Macronutrients

Secondary macronutrients are also essential, even though they are consumed in smaller quantities than N, P, K, that is, primary macronutrients. The secondary macronutrients are Calcium (Ca) Magnesium (Mg) and Sulfur (S).

Calcium

Calcium attaches to the walls of plant tissues, stabilizing the cell wall and favoring cell wall formation. Calcium is also involved in cell growth and development. It improves plant vigor, activating the formation of roots and their growth. Calcium contributes to mineral retention in the soil and to the transportation of such minerals. It neutralizes toxic substances in plants and contributes to seed formation. Calcium stabilizes and regulates several different processes, and a calcium insufficiency causes yellow and brown spots on the leaves. It also slows plant growth in general.

Magnesium

Magnesium constitutes the core of the chlorophyll molecule and is therefore essential for photosynthesis. This makes it an indispensable element for plant development. Magnesium promotes the absorption and transportation of phosphorus. It contributes to the storage of sugars within the plant. Magnesium performs the function of an enzyme activator, and in fact activates more enzymes than any other nutrient. Magnesium deficiencies result in weak stalks, loss of greenness in the oldest leaves, and the appearance of yellow and brown spots, even though the leaves' veins remain green.

Sulfur

Sulfur participates in the formation of chlorophyll. It is necessary for performing photosynthesis and intervenes in protein synthesis and tissue formation. Sulfur is fundamental in the metabolizing of nitrogen, since it improves nitrogen efficiency. Sulfur also improves plant defenses in general. A shortage of sulfur is rare, but when it does occur, the plant becomes lighter in color, taking on a pale green appearance. A general chlorosis is seen, similar to what occurs with a nitrogen deficiency.

Micronutrients

These nutrients are required in lesser amounts as they are toxic at high level. They stimulate plant growth and actively participate in photosynthesis. They show catalytic behavior in plant processes. Their deficiency hampers plant growth, photosynthetic rate reducing crop yield. Deficiency may cause shedding of flowers, poor seed setting and attack of diseases.

Table: Micro-elemental nutrient, their uptake form and their plant mobility.

Nutrient	Uptake form	Mobility in Plant	Mobility in Soil	Typical Range of Soil Content +
Iron	Fe^{2+}, Fe^{3+}	Immobile	Immobile	0.01-21%
Boron	H_3BO_3, BO_3	Immobile	Very mobile	0.9-1000
Copper	Cu^{2+}	Immobile	Immobile	2.5-60
Chlorine	Cl^-	Mobile	Mobile	18-806
Manganese	Mn^{2+}	Immobile	Mobile	<1-18300
Molybdenum	MoO_4^-	Immobile	Somewhat mobile	0.07-5
Zinc	Zn^{2+}	Immobile	Immobile	1.5-2000
Cobalt	Co^{2+}	Immobile	Somewhat immobile	0.3-200
Nickel	Ni^{2+}	Mobile	Somewhat mobile	0.1-1523

+Unit of measurement is mg/kg except where % is used.

Functions of Micronutrients

- Iron: Aids in chlorophyll formation.
- Boron: It regulates nutrients, production of sugar and carbohydrates and also is essential for fruit and seed.
- Copper: It is important for the proper reproductive growth. It abets root metabolism and helps in the consumption of proteins.
- Manganese: It functions with enzyme systems involved in breakdown of carbohydrates and nitrogen metabolism. It regulates the intake and state of oxidation of various elements.
- Chloride: It aids plant carbohydrate metabolism and stimulates enzymatic activity.
- Molybdenum: It helps in the use of nitrogen and in nitrogen fixation as it is constituent of nitrate reductase and nitrogenase enzyme.
- Zinc: It contributes in mechanism for the transformation of carbohydrates and consumption of sugars and also constitutes a part of enzyme systems that regulate plant growth.
- Cobalt: It plays an indispensable role in Vitamin B12 production and thus assists in the process of nitrogen-fixation thus affecting plant growth and metabolism.
- Nickel: It prevents urea accumulation to toxic levels and leaflet tip necrosis. It is essential for the mobilization of stored seed-nitrogen.

Cation Exchange Capacity

It refers to the capacity of soil to exchange cations to enhance the availability of essential trace metals to plants. Mathematically it is defined as number of meq of monovalent cations that can be exchanged per 100g of soil. The value of CEC varies with soil conditions and pH. CEC increases with presence of negatively charged sites on minerals. Humus has high CEC (300-400 meq/100g) while that of organic matter soil (10-30 meq/100g).

Cations are positively charged ions such as calcium (Ca^{2+}), magnesium (Mg^{2+}), and potassium (K^{+}), sodium (Na^{+}) hydrogen (H^{+}), aluminum (Al^{3+}), iron (Fe^{2+}), manganese (Mn^{2+}), zinc (Zn^{2+}) and copper (Cu^{2+}). The capacity of the soil to hold on to these cations called the cation exchange capacity (CEC). These cations are held by the negatively charged clay and organic matter particles in the soil through electrostatic forces (negative soil particles attract the positive cations). The cations on the CEC of the soil particles are easily exchangeable with other cations and as a result, they are plant available. Thus, the CEC of a soil represents the total amount of exchangeable cations that the soil can adsorb.

The cations used by plants in the largest amounts are calcium, magnesium, and potassium. In most soils within humid regions such as in New York, sodium is not present in sufficient quantities to occupy a significant amount of the CEC. However, in dry climates, sodium can occupy an important portion of the CEC. Other cations that can occupy cation exchange sites in New York soils are hydrogen, aluminium, iron and manganese. Cations such as zinc and copper are typically present in the soil in too low a concentration to occupy much of the CEC.

Soils have a CEC primarily because clay particles and organic matter in the soil tends to be negatively charged. New York soils have silicate clay minerals (clay minerals that contain silica). Each silicate clay particle is made up of individual layers or "sheets". If the mineral was pure silica and oxygen (silicaoxide more commonly referred to as quartz), the particle would not have any charge. However, clay minerals common in New York agricultural soils, contain aluminum as well as silica. They have a net negative charge because of the substitution of silica (Si^{4+}) by aluminum (Al^{3+}) in the mineral structure of the clay. This replacement of silica by aluminum in the clay mineral's structure is called isomorphous substitution, and the result is clays with negative surface charge.

Since the soil as a whole does not have electric charge, the negative charge of the clay particles is balanced by the positive charge of the cations in the soil. The negative charges associated with isomorphous substitution are considered permanent, that is, the charges do not change with pH changes.

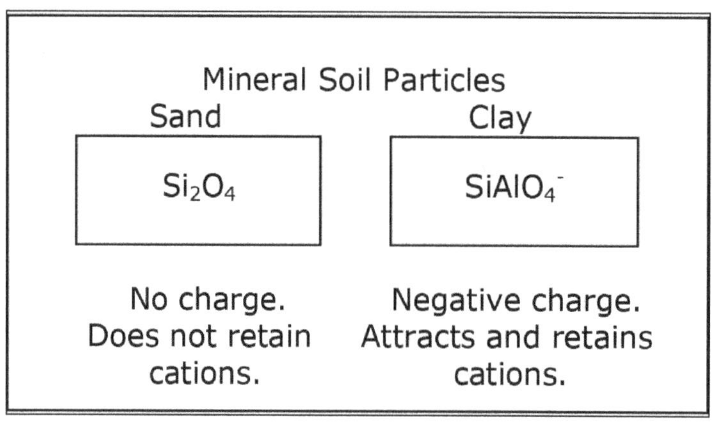

Substitution of silica by aluminum in soil clay particles causes clays to have a negative charge. Because of this negative charge, the soil can hold on to positively charged cations such as calcium (Ca^{2+}), magnesium (Mg^{2+}) and potassium (K^+).

Organic matter can have a 4 to 50 times higher CEC per given weight than clay. The source of negative charge in organic matter is different from that of clay minerals; the dissociation (separation into smaller units) of organic acids causes a net negative charge in soil organic matter, and again this negative charge is balanced by cations in the soil. Because organic acid dissociation depends on the soil pH, the CEC associated with soil organic matter is called pH-dependent CEC. This means that the actual CEC of the soil will depend on the pH of the soil. Given the same amount and type of organic matter, a neutral soil (pH ~7) will have a higher CEC than a soil with e.g. pH 5, or in other words, the CEC of a soil with pH-dependent charge will increase with an increase in pH.

Units

The CEC of a soil is expressed in $cmol_c/kg$ (centimol positive charge per kg of soil) or meq/100 g (milli-equivalents per 100 grams of soil). Both expressions are numerically identical (10 $cmol_c/kg$ = 10 meq/100 g).

Implications

- The higher the CEC the more clay or organic matter present in the soil. This usually means that high CEC (clay) soils have a greater water holding capacity than low CEC (sandy) soils.
- Low CEC soils are more likely to develop potassium and magnesium (and other cation) deficiencies, while high CEC soils are less susceptible to leaching losses of these cations. So, for sandy soils, a large one-time addition of cations e.g. potassium can lead to large leaching losses (soil isn't able to hold on to the excess K). More frequent additions of smaller amounts are better.
- The lower the CEC, the faster the soil pH will decrease with time. So, sandy soils need to be limed more often than clay soils.
- The higher the CEC, the larger the quantity of lime that must be added to increase the soil pH; sandy soils need less lime than clay soils to increase the pH to desired levels.

Anion Exchange Capacity

It is defined as the measurement of the positive charges in soils being affected by the amount of negative charges which a soil can absorb. The value of AEC is directly proportional to change in pH while in case of CEC, reverse trend is seen. The relative strength of adsorption of cations and anions can be illustrated by Lyotrophilic series as follows:

Cations: $Al^{3+} > H^{+} > Ca^{2+} > Mg^{2+} > K^{+} = NH_4^{+} > Na^{+}$

Anions: $H_2PO_4^{-} > SO_4^{2-} > NO_3^{-} > Cl$

Anion-exchange capacity has been defined as the capacity of a soil to adsorb or release anions under normal soil conditions. The anion exchange is known to increase with adsorption of cations like zinc as $Zn(OH)^{+}$ or $ZnCl^{+}$ or both. The concentration of anions in the soil solution is affected by the solid phase. The overall negative charge of the exchange complex causes anions to be negatively adsorbed.

Under certain conditions hydrous oxides of iron and aluminium, allophone and even kaolinite show evidence of having positive charges on their crystal surfaces. These charges arise from the following two sources:

- The protonation or adding of hydrogen ions to hydroxyl groups on the edge of these minerals.

$$\underset{\substack{\text{Surface of mineral} \\ \text{(High pH)}}}{Al{-}O...H + H^{+}} \xrightleftharpoons{\text{Protonation reaction}} \underset{\substack{\text{Surface of mineral} \\ \text{(Low pH)}}}{Al{-}OH_2^{+}}$$

At high pH values (left hand side) the hydrogen ion tends to dissociate from the oxygen, leaving a negative charge on the surface. As the pH is lowered (right hand side), an additional hydrogen ion associates with the hydroxyl, leaving a net positive charge.

- The exchange of the hydroxyl (-OH) groups for other anions, e.g. phosphates and other similar anions (Anion exchange).

$$>\text{Al—OH} + H_2PO_4^- \rightleftharpoons >\text{Al—}H_2PO_4 + OH^-$$

Surface of mineral Surface of mineral

This reaction also occurs primarily at low pH values, the negative hydroxyl ion being replaced from positively charged aluminium ions in the crystal.

Factor Affecting Anion Exchange

- Soil pH: Contrary to cation exchange, the capacity for retaining anions increases with a decrease in soil pH.
- Valency of Ions: Anion exchange increases with the valence of the anion but decreases with increasing valence of the cation.
- Type and Amount of Clay Colloid: Anion exchange is much greater in soils high in 1: 1 type clay minerals and those containing hydrous oxides of iron and aluminium then that of soils containing high amounts of 2: 1 type of clay minerals. Clay minerals in the montmorillonite group of expansible layer silicates usually have anion exchange capacities of the less than 5 me/100 g. On the other hand, kaolinites can have an anion exchange capacity as high as 43 me/100 g at an acidic equilibrium pH of 4.7.
- Salt Concentration: Anion exchange generally decreases with increasing salt concentration. Addition of salts suppresses the unequal ion distribution. In general the relative order of anion exchange is $OH^- > H_2PO_4^- > SO_4^{2-} > NO_3^-$.

Importance of Anion Exchange

Among various anions, the exchangeable phosphate is of most importance since sulphates or nitrates are not retained in sufficient quantities. So the phenomenon of anion exchange is most important for the release of fixed phosphate in the soil and thereby increases its availability to plants. The anion exchange is carried out mainly by the replacement of OH ions of the clay minerals. The reaction is as follows:

$$\boxed{\text{Clay micelle}}\,OH + H_2PO_4^- \rightleftharpoons \boxed{\text{Clay micelle}}\,H_2PO_4 + OH^-$$

Adsorbed

Such type of phosphate adsorption renders phosphate unavailable to plants. Since the anion exchange is reversible phenomenon a part of the adsorbed phosphate ions again become readily available when soils are dominated with OH ions i.e. means under high pH conditions (may be induced with liming).

The origin of OH ions is from silicate clay minerals and hydrous oxides of iron and aluminium that are present in the soil. Large quantities of soluble phosphate fixed by these minerals as follows:

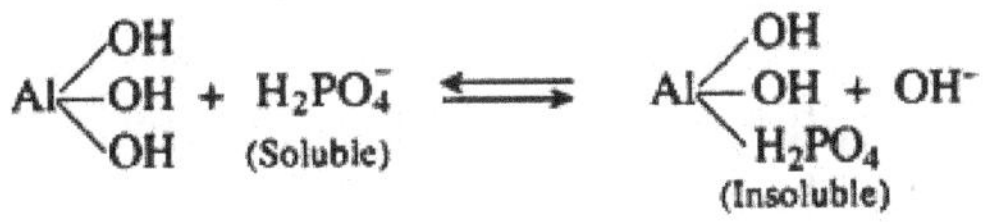

In this reaction takes place under slightly acidic soil conditions, the exchange phenomenon is reversible, and soluble phosphate is further released when hydroxy phosphate compound (insoluble) comes in contact with OH ions.

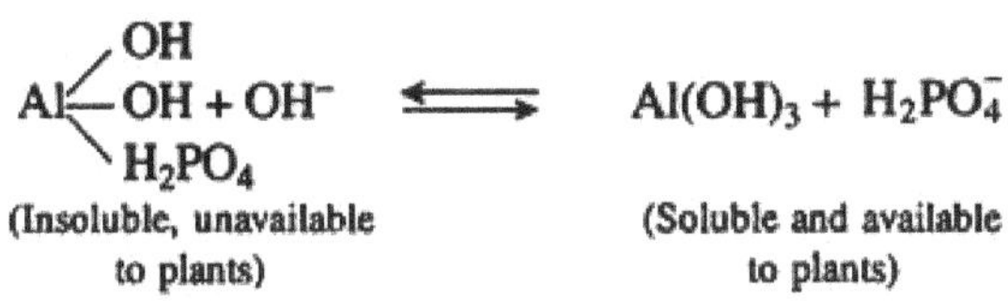

In this reaction occurs under moderate to strong acid conditions, the phosphate ions are irreversibly fixed and not available to plants.

Soil Salinity

Salinity is one of the most brutal environmental factors limiting the productivity of crop plants because most of the crop plants are sensitive to salinity caused by high concentrations of salts in the soil, and the area of land affected by it is increasing day by day. For all important crops, average yields are only a fraction – somewhere between 20% and 50% of record yields; these losses are mostly due to drought and high soil salinity, environmental conditions which will worsen in many regions because of global climate change. A wide range of adaptations and mitigation strategies are required to cope with such impacts. Efficient resource management and crop/livestock improvement for evolving better breeds can help to overcome salinity stress. However, such strategies being long drawn and cost intensive, there is a need to develop simple and low cost biological methods for salinity stress management, which can be used on short term basis. Microorganisms could play a significant role in this respect, if we exploit their unique properties such as tolerance to saline conditions, genetic diversity, synthesis of compatible solutes, production of plant growth promoting hormones, bio-control potential, and their interaction with crop plants.

A saline soil is generally defined as one in which the electrical conductivity (EC) of the saturation extract (EC_e) in the root zone exceeds 4 dS m^{-1} (approximately 40 mM NaCl) at 25 °C and has an exchangeable sodium of 15%. The yield of most crop plants is reduced at this EC_e, though many crops exhibit yield reduction at lower EC_es. It has been estimated that worldwide 20% of total cultivated and 33% of irrigated agricultural lands are afflicted by high salinity. Furthermore, the salinized areas are increasing at a rate of 10% annually for various reasons, including low precipitation, high surface evaporation, weathering of native rocks, irrigation with saline water, and poor cultural practices.

Water and soil management practices have facilitated agricultural production on soil marginalized by salinity but an additional gain from these approaches seems problematic. Impacted soils are a major limiting production factor worldwide for every major crop. A significant increase (an estimated 50%) in grain yields of major crop plants such as rice (*Oryza sativa* L.), wheat (*Triticum aestivum* L.) and maize (*Zea mays* L.) is required to fulfill the food supply requirements for the projected population by 2050. The urgency of feeding the world's growing population while combating soil pollution, salinization, and desertification has given plant and soil productivity research vital importance. Under such circumstances, it requires suitable biotechnology not only to improve crop productivity but also to improve soil health through interactions of plant roots and soil microorganisms.

Salt stressed soils are known to suppress the growth of plants. Plants in their natural environment are colonized both by endocellular and intracellular microorganisms. Rhizosphere microorganisms, particularly beneficial bacteria and fungi, can improve plant performance under stress environments and, consequently, enhance yield both directly and indirectly. Some plant growth-promoting rhizobacteria (PGPR) may exert a direct stimulation on plant growth and development by providing plants with fixed nitrogen, phytohormones, iron that has been sequestered by bacterial siderophores, and soluble phosphate. Others do this indirectly by protecting the plant against soil-borne diseases, most of which are caused by pathogenic fungi. The problem of soil salinization is a scourge for agricultural productivity worldwide. Crops grown on saline soils suffer on an account of high osmotic stress, nutritional disorders and toxicities, poor soil physical conditions and reduced crop productivity.

Problem of Soil Salinization

Soil salinity is an enormous problem for agriculture under irrigation. In the hot and dry regions of the world the soils are frequently saline with low agricultural potential. In these areas most crops are grown under irrigation, and to exacerbate the problem, inadequate irrigation management leads to secondary salinization that affects 20% of irrigated land worldwide. Irrigated agriculture is a major human activity, which often leads to secondary salinization of land and water resources in arid and semi-arid conditions. Salts in the soil occur as ions (electrically charged forms of atoms or compounds). Ions are released from weathering minerals in the soil. They may also be applied through irrigation water or as fertilizers, or sometimes migrate upward in the soil from shallow groundwater. When precipitation is insufficient to leach ions from the soil profile, salts accumulate in the soil resulting soil salinity. All soils contain some water-soluble salts. Plants absorb essential nutrients in the form of soluble salts, but excessive accumulation strongly suppresses the plant growth. During the last century, physical, chemical and/or biological land degradation processes have resulted in serious consequences to global natural resources (e.g. compaction, inorganic/organic contamination, and diminished microbial activity/diversity). The area under the affected soils continues to increase each year due to introduction of irrigation in new areas.

Impact of Salinity on Plants

Agricultural crops exhibit a spectrum of responses under salt stress. Salinity not only decreases the agricultural production of most crops, but also, effects soil physicochemical properties, and ecological balance of the area. The impacts of salinity include—low agricultural productivity, low economic returns and soil erosions,. Salinity effects are the results of complex interactions among morphological, physiological, and biochemical processes including seed germination, plant growth, and water and nutrient uptake. Salinity affects almost all aspects of plant development including: germination, vegetative growth and reproductive development. Soil salinity imposes ion toxicity, osmotic stress, nutrient (N, Ca, K, P, Fe, Zn) deficiency and oxidative stress on plants, and thus limits water uptake from soil.

Soil salinity significantly reduces plant phosphorus (P) uptake because phosphate ions precipitate with Ca ions. Some elements, such as sodium, chlorine, and boron, have specific toxic effects on plants. Excessive accumulation of sodium in cell walls can rapidly lead to osmotic stress and cell death. Plants sensitive to these elements may be affected at relatively low salt concentrations if the soil contains enough of the toxic element. Because many salts are also plant nutrients, high salt levels in the soil can upset the nutrient balance in the plant or interfere with the uptake of some nutrients. Salinity also affects photosynthesis mainly through a reduction in leaf area, chlorophyll content and stomatal conductance, and to a lesser extent through a decrease in photosystem II efficiency. Salinity adversely affects reproductive development by inhabiting microsporogenesis and stamen filament elongation, enhancing programed cell death in some tissue types, ovule abortion and senescence of fertilized embryos. The saline growth medium causes many adverse effects on plant growth, due to a low osmotic potential of soil solution (osmotic stress), specific ion effects (salt stress), nutritional imbalances, or a combination of these factors. All these factors cause adverse effects on plant growth and development at physiological and biochemical levels, and at the molecular level.

In order to assess the tolerance of plants to salinity stress, growth or survival of the plant is measured because it integrates the up- or down-regulation of many physiological mechanisms occurring within the plant. Osmotic balance is essential for plants growing in saline medium. Failure of this balance results in loss of turgidity, cell dehydration and ultimately, death of cells. On the other hand, adverse effects of salinity on plant growth may also result from impairment of the supply of photosynthetic assimilates or hormones to the growing tissues. Ion toxicity is the result of replacement of K^+ by Na^+ in biochemical reactions, and Na^+ and Cl^- induced conformational changes in proteins. For several enzymes, K^+ acts as cofactor and cannot be substituted by Na^+. High K^+ concentration is also required for binding tRNA to ribosomes and thus protein synthesis. Ion toxicity and osmotic stress cause metabolic imbalance, which in turn leads to oxidative stress. The adverse effects of salinity on plant development are more profound during the reproductive phase. Wheat plants stressed at 100–175 mM NaCl showed a significant reduction in spikelets per spike, delayed spike emergence and reduced fertility, which results in poor grain yields. However, Na^+ and Cl^- concentrations in the shoot apex of these wheat plants were below 50 and 30 mM, respectively, which is too low to limit metabolic reactions.

Hence, the adverse effects of salinity may be attributed to the salt-stress effect on the cell cycle and differentiation. Salinity arrests the cell cycle transiently by reducing the expression and activity of cyclins and cyclin-dependent kinases that results in fewer cells in the meristem, thus limiting growth. The activity of cyclin-dependent kinase is diminished also by post-translational inhibition during salt stress. Recent reports also show that salinity adversely affects plant growth and development, hindering seed germination, seedling growth, enzyme activity, DNA, RNA, protein synthesis and mitosis.

Amelioration of Salinity

Salinization can be restricted by leaching of salt from root zone, changed farm management practices and use of salt tolerant plants. Irrigated agriculture can be sustained by better irrigation practices such as adoption of partial root zone drying methodology, and drip or micro-jet irrigation to optimize use of water. The spread of dry land salinity can be contained by reducing the amount of water passing beyond the roots. This can be done by re-introducing deep rooted perennial plants that continue to grow and use water during the seasons that do not support annual crop plants. This may restore the balance between rainfall and water use, thus preventing rising water tables and the movement of salt to the soil surface. Farming systems can change to incorporate perennials in rotation with annual crops (phase farming), in mixed plantings (alley farming, intercropping), or in site-specific plantings (precision farming). Although the use of these approaches to sustainable management can ameliorate yield reduction under salinity stress, implementation is often limited because of cost and availability of good water quality or water resource. Evolving efficient, low cost, easily adaptable methods for the abiotic stress management is a major challenge. Worldwide, extensive research is being carried out, to develop strategies to cope with abiotic stresses, through development of salt and drought tolerant varieties, shifting the crop calendars, resource management practices etc. as shown in Figure.

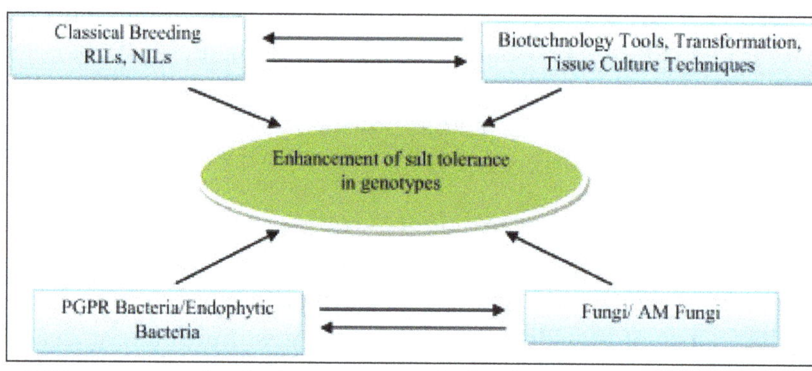

Different approaches for improvement of salt tolerance in agricultural crops.

Use of Salt Tolerant Crops and Transgenics

Using the salt-tolerant crops is one of the most important strategies to solve the problem of salinity. Tolerance will be required for the "de-watering" species, but also for the annual crops to follow, as salt will be left in the soil when the water table is lowered. Salt tolerance in crops will also allow the more effective use of poor quality irrigation water. To increase the plant salt-tolerance, there is a need for understanding the mechanisms of salt limitation on plant growth and the mechanism of salt tolerance at the whole-plant, organelle, and molecular levels. Under saline conditions, there is a change in the pattern of gene expression, and both qualitative and quantitative changes in protein synthesis. Although it is generally agreed that salt stress brings about quantitative changes in protein synthesis, there is some controversy as to whether salinity activates specialized genes that are involved in salt stress. Salt tolerance does not appear to be conferred by unique gene(s). When a plant is subjected to abiotic stress, a number of genes are turned on, resulting in increased levels of several metabolites and proteins, some of which may be responsible for conferring a certain degree of protection to these stresses. Efforts to improve crop performance by transgenic approach under environmental stresses have not been that fruitful because the fundamental mechanisms of stress tolerance in plants remain to be completely understood.

Development of salt-tolerant crops has been a major objective of plant breeding programs for decades in order to maintain crop productivity in semiarid and saline lands. Although several salt-tolerant varieties have been released, the overall progress of traditional breeding has been slow and has not been successful as only few major determinant genetic traits of salt tolerance have been identified. 25 years ago Epstein et al. described the technical and biological constraints to solving the problem of salinity. Although there has been some success with technical solutions to the problem, the biological solutions have been more difficult to develop because a pre-requisite for the development of salt tolerant crops is the identification of key genetic determinants of stress tolerance. The existence of salt-tolerant plants (halophytes) and differences in salt tolerance between genotypes within salt-sensitive plant species (glycophytes) indicates that there is a genetic basis to salt response. Although a lot of approaches have been done for development of salt tolerant plants by transgenics complete success is not achieved yet. The assessment of salt tolerance in transgenic experiments has been mostly carried out using a limited number of seedlings or mature plants in laboratory experiments. In most of the cases, the experiments were carried out in greenhouse conditions where the plants were not exposed to those conditions that prevail in high-salinity soils (e.g. alkaline soil pH, high diurnal temperatures, low humidity, and presence of other sodic salts and elevated concentrations of selenium and/or boron).

The salt tolerance of the plants in the field needs to be evaluated and, more importantly, salt tolerance needs to be evaluated as a function of yield. The evaluation of field performance under salt stress is difficult because of the variability of salt levels in field conditions and the potential for interactions with other environmental factors, including soil fertility, temperature, light intensity and water loss due to transpiration. Evaluating tolerance is also made more complex because of variation in sensitivity to salt during the life cycle. For example, in rice, grain yield is much more affected by salinity than in vegetative growth. In tomato, the ability of the plants to germinate under conditions of high salinity is not always correlated with the ability of the plant to grow under salt stress because both are controlled by different mechanisms, although some genotypes might display similar tolerance at germination and during vegetative growth. Therefore, the assessment of stress tolerance in the laboratory often has little correlation to tolerance in the field. Although there have been many successes in developing stress-tolerant transgenics in model plants such as tobacco, Arabidopsis or rice, there is an urgent need to test these successes in other crops. There are several technical and financial challenges associated with transforming many of the crop plants, particularly the monocots. First, transformation of any monocot other than rice is still not routine and to develop a series of independent homozygous lines is costly, both in terms of money and time. Second, the stress tolerance screens will need to include a field component because many of the stress tolerance assays used by basic researchers involve using nutrient-rich media (which in some cases include sucrose). This type of screen is unlikely to have a relationship to field performance. Third, because saline soils are often complex and can include NaCl, $CaCl_2$, $CaSO_4$, Na_2SO_4, high boron concentrations and alkaline pH, plants that show particular promise will eventually have to be tested in all these environments.

Microbes: Abiotic Stress Alleviation Tool in Crops

Several strategies have been developed in order to decrease the toxic effects caused by high salinity on plant growth, including plant genetic engineering, and recently the use of plant growth-promoting bacteria (PGPB). The role of microorganisms in plant growth promotion, nutrient management and disease control is well known and well established. These beneficial microorganisms colonize the rhizosphere/endorhizosphere of plants and promote growth of the plants through various direct and indirect mechanisms.

The term Induced Systemic Tolerance (IST) has been proposed for PGPR-induced physical and chemical changes that result in enhanced tolerance to abiotic stress. PGPR facilitate plant growth indirectly by reducing plant pathogens, or directly by facilitating the nutrient uptake through phytohormone production (e.g. auxin, cytokinin and gibberellins), by enzymatic lowering of plant ethylene levels and/or by production of siderophores. It has been demonstrated that inoculations with AM (arbuscular mycorrhizal) fungi improves plant growth under salt stress. Kohler et al., 2006 demonstrated the beneficial effect of PGPR Pseudomonas mendocina strains on stabilization of soil aggregate. The three PGPR isolates P. alcaligenes PsA15, Bacillus polymyxa BcP26 and Mycobacterium phlei MbP18 were able to tolerate high temperatures and salt concentrations and thus confer on them potential competitive advantage to survive in arid and saline soils such as calcisol. Kohler et al., 2009 investigated the influence of inoculation with a PGPR, P. mendocina, alone or in combination with an AM fungus, Glomus intraradices or G. mosseae on growth and nutrient uptake and other physiological activities of Lactuca sativa affected by salt stress.

The plants inoculated with P. mendocina had significantly greater shoot biomass than the controls and it is suggested that inoculation with selected PGPR could be an effective tool for alleviating salinity stress in salt sensitive plants. Bacteria isolated from different stressed habitats possess stress tolerance capacity along with the plant growth-promoting traits and therefore are potential candidates for seed bacterization. When inoculated with these isolates, plants show enhanced root and shoot length, biomass, and biochemical levels such as chlorophyll, carotenoids, and protein. Investigations on interaction of PGPR with other microbes and their effect on the physiological response of crop plants under different soil salinity regimes are still in incipient stage. Inoculations with selected PGPR and other microbes could serve as the potential tool for alleviating salinity stress in salt sensitive crops. Therefore, an extensive investigation is needed in this area, and the use of PGPR and other symbiotic microorganisms, can be useful in developing strategies to facilitate sustainable agriculture in saline soils.

Alleviation of Abiotic Stress in Plants By Rhizospheric Bacteria

Besides developing mechanisms for stress tolerance, microorganisms can also impart some degree of tolerance to plants towards abiotic stresses like drought, chilling injury, salinity, metal toxicity and high temperature. In the last decade, bacteria belonging to different genera including Rhizobium, Bacillus, Pseudomonas, Pantoea, Paenibacillus, Burkholderia, Achromobacter, Azosirillum, Microbacterium, Methylobacterium, Variovorax, Enterobacter etc. have been reported to provide tolerance to host plants under different abiotic stress environments. Use of these microorganisms per se can alleviate stresses in agriculture thus opening a new and emerging application of microorganisms. Microbial elicited stress tolerance in plants may be due to a variety of mechanisms proposed from time to time based on studies done. Production of indole acetic acid, gibberellins and some unknown determinants by PGPR, results in increased root length, root surface area and number of root tips, leading to an enhanced uptake of nutrients thereby improving plant health under stress conditions. Plant growth promoting bacteria have been found to improve growth of tomato, pepper, canola, bean and lettuce under saline conditions.

Some PGPR strains produce cytokinin and antioxidants, which result in abscisic acid (ABA) accumulation and degradation of reactive oxygen species. High activities of antioxidant enzymes are linked with oxidative stress tolerance. Another PGPR strain, Achromobacter piechaudii ARV8 which produced 1-aminocyclopropane-1-carboxylate (ACC) deaminase, conferred IST against drought and salt in pepper and tomato. Many aspects of plant life are regulated by ethylene levels and the biosynthesis of ethylene is subjected to tight regulation, involving transcriptional and post-transcriptional factors regulated by environmental cues, including biotic and abiotic stresses. Under stress conditions, the plant hormone ethylene endogenously regulates plant homoeostasis and results in reduced root and shoot growth. In the presence of ACC deaminase producing bacteria, plant ACC is sequestered and degraded by bacterial cells to supply nitrogen and energy.

Furthermore, by removing ACC, the bacteria reduce the deleterious effect of ethylene, ameliorating stress and promoting plant growth. The complex and dynamic interactions among microorganisms, roots, soil and water in the rhizosphere induce changes in physicochemical and structural properties of the soil. Microbial polysaccharides can bind soil particles to form microaggregates and macroaggregates. Plant roots and fungal hyphae fit in the pores between microaggregates and thus stabilize macroaggregates. Plants treated with Exo-poly saccharides (EPS) producing bacteria display increased resistance to water and salinity stress due to improved soil structure. EPS can also bind to cations including $Na+$ thus making it unavailable to plants under saline conditions. Chen et al., 2007 correlated proline accumulation with drought and salt tolerance in plants. Introduction of proBA genes derived from B. subtilis into A. thaliana resulted in production of higher levels of free proline resulting in increased tolerance to osmotic stress in the transgenic plants. Increased production of proline along with decreased electrolyte leakage, maintenance of relative water content of leaves and selective uptake of K ions resulted in salt tolerance in Zea mays coinoculated with Rhizobium and Pseudomonas Rhizobacteria inhabiting the sites exposed to frequent stress conditions, are likely to be more adaptive or tolerant and may serve as better plant growth promoters under stressful conditions. Moreover Yao et al., 2010 reported that inoculation with P. putida Rs 198 promoted cotton growth and germination under conditions of salt stress. Tank and Saraf showed that PGPRs which are able to solubilize phosphate, produce phytohormones and siderophores in salt condition promote growth of tomato plants under 2% NaCl stress.

In a study carried out by Naz et al., 2009, it was shown that strains isolated from Khewra salt range of Pakistan exhibited their tolerance when tested on saline media simulated by rhizosphere soil filtrate. Noteworthy, the isolates produced ABA in a concentration much higher than that of previous reports. Furthermore production of proline, shoot/root length, and dry weight was also higher in soybean plants inoculated with these isolates under induced salt stress. Likewise Upadhyay et al., 2011 studied the impact of PGPR inoculation on growth and antioxidant status of wheat under saline conditions and reported that co-inoculation with B. subtilis and Arthrobacter sp. could alleviate the adverse effects of soil salinity on wheat growth with an increase in dry biomass, total soluble sugars and proline content. Jha et al., 2011 reported that P. pseudoalcaligenes, an endophytic bacterium in combination with a rhizospheric B. pumilus in paddy was able to protect the plant from abiotic stress by induction of osmoprotectant and antioxidant proteins than by the rhizospheric or endophytic bacteria alone at early stages of growth. Plants inoculated with endophytic bacterium P. pseudoalcaligenes showed a significantly higher concentration of glycine betaine-like quaternary compounds and higher shoot biomass at lower salinity levels. While at higher salinity levels, a mixture of both P. pseudoalcaligenes and B. pumilus showed better response against the adverse effects of salinity. Nia et al., 2012 studied the effect of inoculation of Azospirillum strains isolated from saline or non-saline soil on yield and yield components of wheat in salinity and they observed that inoculation with the two isolates increased salinity tolerance of wheat plants; the saline-adapted isolate significantly increased shoot dry weight and grain yield under severe water salinity. The component of grain yield most affected by inoculation was grains per plant. Plants inoculated with saline-adapted Azospirillum strains had higher N concentrations at all water salinity levels.

The plant growth promoting activity of an auxin and siderophore producing isolate of Streptomyces under saline soil conditions and reported increases in growth and development of wheat plant. They observed significant increases in germination rate, percentage and uniformity, shoot length and dry weight compared to the control. Applying the bacterial inocula increased the concentration of N, P, Fe and Mn in wheat shoots grown in normal and saline soil and thus concluded that Streptomyces isolate has potential to be utilized as biofertilizers in saline soils. More recently Ramadoss et al., 2013 studied the effect of five plant growth promoting halotolerant bacteria on wheat growth and found that inoculation of those halotolerant bacterial strains to ameliorate salt stress (80, 160 and 320 mM) in wheat seedlings produced an increase in root length of 71.7% in comparison with uninoculated positive controls. In particular, Hallobacillus sp. and B. halodenitrificans showed more than 90% increase in root elongation and 17.4% increase in dry weight when compared to uninoculated wheat seedlings at 320 mM NaCl stress indicating a significant reduction of the deleterious effects of NaCl. These results indicate that halotolerant bacteria isolated from saline environments have potential to enhance plant growth under saline stress through direct or indirect mechanisms and would be most appropriate as bioinoculants under such conditions. The isolation of indigenous microorganisms from the stress affected soils and screening on the basis of their stress tolerance and PGP traits may be useful in the rapid selection of efficient strains that could be used as bioinoculants for stressed crops.

Soil pH

Soil pH or soil reaction is an indication of the acidity or alkalinity of soil and is measured in pH units. Soil pH is defined as the negative logarithm of the hydrogen ion concentration. The pH scale goes from 0 to 14 with pH 7 as the neutral point. As the amount of hydrogen ions in the soil increases the soil pH decreases thus becoming more acidic. From pH 7 to 0 the soil is increasingly more acidic and from pH 7 to 14 the soil is increasingly more alkaline or basic.

Descriptive terms commonly associated with certain ranges in soil pH are:

- Extremely acid: < than 4.5; lemon = 2.5; vinegar = 3.0; stomach acid = 2.0; soda = 2 – 4
- Very strongly acid: 4.5 – 5.0; beer = 4.5 – 5.0; tomatoes = 4.5
- Strongly acid: 5.1 – 5.5; carrots = 5.0; asparagus = 5.5; boric acid = 5.2; cabbage = 5.3
- Moderately acid: 5.6 – 6.0; potatoes = 5.6

- Slightly acid: 6.1 – 6.5; salmon = 6.2; cow's milk = 6.5
- Neutral: 6.6 – 7.3; saliva = 6.6 – 7.3; blood = 7.3; shrimp = 7.0
- Slightly alkaline: 7.4 – 7.8; eggs = 7.6 – 7.8
- Moderately alkaline: 7.9 – 8.4; sea water = 8.2; sodium bicarbonate = 8.4
- Strongly alkaline: 8.5 – 9.0; borax = 9.0
- Very strongly alkaline: > than 9.1; milk of magnesia = 10.5, ammonia = 11.1; lime = 12

Measuring Soil pH

Soil pH provides various clues about soil properties and is easily determined. The most accurate method of determining soil pH is by a pH meter. A second method which is simple and easy but less accurate then using a pH meter, consists of using certain indicators or dyes.

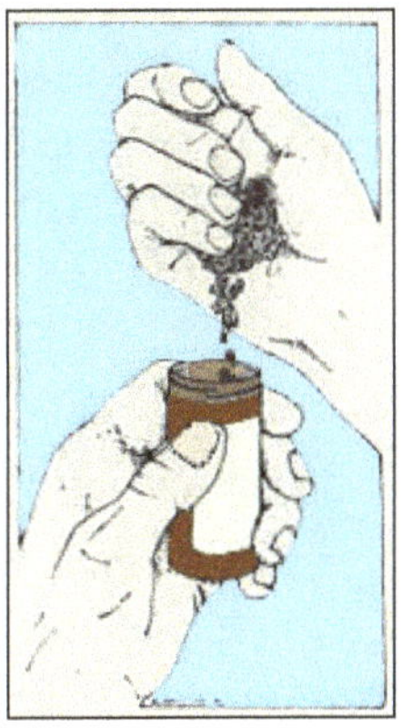

Many dyes change color with an increase or decrease of pH making it possible to estimate soil pH. In making a pH determination on soil, the sample is saturated with the dye for a few minutes and the color observed. This method is accurate enough for most purposes. Kits (pH) containing the necessary chemicals and color charts are available from garden stores.

There may be considerable variation in the soil pH from one spot in a field or lawn to another. To determine the average soil pH of a field or lawn it is necessary to collect soil from several locations and combine into one sample.

pH Affects Nutrients, Minerals and Growth

The effect of soil pH is great on the solubility of minerals or nutrients. Fourteen of the seventeen essential plant nutrients are obtained from the soil. Before a nutrient can be used by plants it must be dissolved in the soil solution. Most minerals and nutrients are more soluble or available in acid soils than in neutral or slightly alkaline soils.

Phosphorus is never readily soluble in the soil but is most available in soil with a pH range centered around 6.5. Extremely and strongly acid soils (pH 4.0-5.0) can have high concentrations of soluble aluminum, iron and manganese which may be toxic to the growth of some plants. A pH range of approximately 6 to 7 promotes the most ready availability of plant nutrients.

But some plants, such as azaleas, rhododendrons, blueberries, white potatoes and conifer trees, tolerate strong acid soils and grow well. Also, some plants do well only in slightly acid to moderately alkaline soils. However, a slightly alkaline (pH 7.4-7.8) or higher pH soil can cause a problem with the availability of iron to pin oak and a few other trees in Central New York causing chlorosis of the leaves which will put the tree under stress leading to tree decline and eventual mortality.

<table>
<tr><th>Soil pH</th><th>Plant Growth</th></tr>
<tr><td>> 8.3</td><td>Too alkaline for most plants</td></tr>
<tr><td>7.5</td><td>Iron availability becomes a problem on alkaline soils</td></tr>
<tr><td>7.2</td><td rowspan="4">6.8 to 7.2 – near neutral
6.0 to 7.5 – acceptable for most plants</td></tr>
<tr><td>7.0</td></tr>
<tr><td>6.8</td></tr>
<tr><td>6.0</td></tr>
<tr><td>5.5</td><td>Reduced soil microbial activity</td></tr>
<tr><td>< 4.6</td><td>Too acid for most plants</td></tr>
</table>

The soil pH can also influence plant growth by its effect on activity of beneficial microorganisms Bacteria that decompose soil organic matter are hindered in strong acid soils. This prevents organic matter from breaking down, resulting in an accumulation of organic matter and the tie up of nutrients, particularly nitrogen, that are held in the organic matter.

Changes in Soil pH

Soils tend to become acidic as a result of: (1) rainwater leaching away basic ions (calcium, magnesium, potassium and sodium); (2) carbon dioxide from decomposing organic matter and root respiration dissolving in soil water to form a weak organic acid; (3) formation of strong organic and inorganic acids, such as nitric and sulfuric acid, from decaying organic matter and oxidation of ammonium and sulfur fertilizers. Strongly acid soils are usually the result of the action of these strong organic and inorganic acids.

Lime is usually added to acid soils to increase soil pH. The addition of lime not only replaces hydrogen ions and raises soil pH, thereby eliminating most major problems associated with acid soils but it also provides two nutrients, calcium and magnesium to the soil. Lime also makes phosphorus that is added to the soil more available for plant growth and increases the availability of nitrogen by hastening the decomposition of organic matter. Liming materials are relatively inexpensive, comparatively mild to handle and leave no objectionable residues in the soil.

Some common liming materials are: (1) Calcic limestone which is ground limestone; (2) Dolomitic limestone from ground limestone high in magnesium; and (3) Miscellaneous sources such as wood ashes. The amount of lime to apply to correct a soil acidity problem is affected by a number of factors, including soil pH, texture (amount of sand, silt and clay), structure, and amount of organic matter. In addition to soil variables the crops or plants to be grown influence the amount of lime needed.

Adjustment of Soil Acidity

- If the soil becomes too acidic, it can be refurbished by liming:

  $\{Soil\} (H^+)_2 + CaCO_3 \quad \{Soil\} Ca^{2+} + CO_2 + H_2O$

- If soil becomes too alkaline (in areas receiving scanty rainfall), it can be treated with aluminum or iron sulfate, releases acid on hydrolysis:

  $2Fe^{3+} + 3SO_4^{2-} + 6H_2O \quad 2Fe(OH)_3 + 6H^+ + 3SO_4^{2-}$

- Sulphur is added to soils is oxidized by bacterially mediated reactions to sulphuric acid:

  $S + O_2 + H_2O \quad 2H^+ + SO_4^{2-}$

4

Nutrient Cycle in Soil

Nutrient cycling in soil refers to putting back the nutrients of soil into biological circulation by liberating carbon and nutrients from complex materials through decomposition. This chapter discusses in detail ion absorption by plant roots and carbon, nitrogen, phosphorus and potassium cycle in soil. The chapter closely examines the key concepts of soil nutrient cycle to provide an extensive understanding of the subject.

Ion Absorption by Plant Roots

Ion Absorption in Plants– Passive and Active Uptake

Passive Uptake of Ion Absorption

Non-Mediated Passive Uptake

Numerous investigators have demonstrated non-metabolic or passive uptake of ions due to the fact that when a plant cell or tissue is transferred from a low-salt concentration medium to a relatively high-salt concentration medium, there is an initial uptake of ions due to diffusion.

This initial rapid uptake is temperature independent and remains unaffected by the application of metabolic inhibitors. Passive absorption includes theories of diffusion, ion exchange, Donnan equilibrium, mass flow, etc.

(a) Diffusion:

If the above cell or tissue is again returned to a lower-salt solution, some of the absorbed ions will diffuse out into the external medium. This process of free diffusion of solutes follows the simple laws of diffusion. This free diffusion means free movement of ions in and out of the tissue.

The part of the tissue for free diffusion shows equilibrium with the external medium. This part of a cell or tissue is referred to as outer space. The term apparent free space is also used to denote the apparent volume accommodating the freely diffused ions.

(b) Ion Exchange:

In the ion exchange mechanism the ions within the cells are exchanged for the ions of equivalent charge of the external solution. If a solution of a dissociable salt ($A+B-$) is separated from distilled water by a membrane which is permeable to both ions diffusion will occur until the concentration of salt on both sides are equal. If the membrane is impermeable to either cation or anion, there will be no ion movement.

Now, if two solutions containing different salts ($A_1^+B_1^-$ and $A_2^+B_2^-$) are separated from one another by either a cation-permeable/anion-impermeable or cation-impermeable/anion- permeable membrane, one of the two ion species in each case is free to move across in exchange for an ion of same charge.

At equilibrium the ratio $[A_1^+]/[A_2^+]$ (when cation is permeable) or $[B_1^-]/[B_2^-]$ (when anion is permeable) is equal on the two sides.

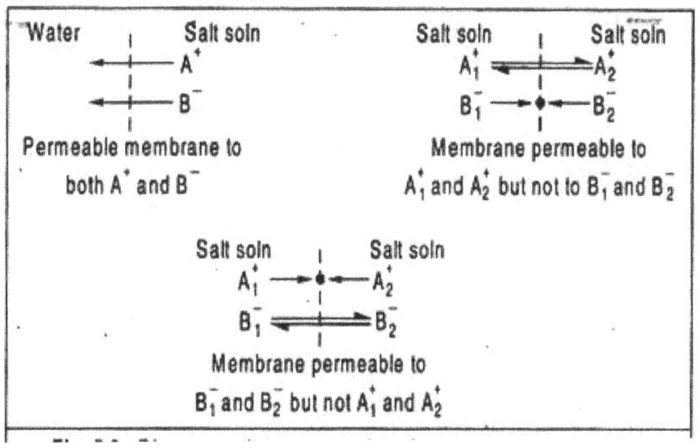

The total concentration of salts on either side is not affected by the exchanges of ions. Exchange involves equivalent electrical charge, so that two univalent ions are exchanged for one bivalent, three for one trivalent ion, and so on.

With the help of radioactive isotopes it is possible to calculate the rates of exchange of ions between the two solutions kept separated by an ion exchange membrane from measurements of changes in radioactivity with time on either side.

(c) Donnan Equilibrium:

The Donnan equilibrium theory accounts for the effect of fixed or non-diffusible ions and explains the co-operation of both electrical as well as diffusion phenomena. It is a complex ion-exchange system in which the membrane is impermeable to certain ions called fixed ions (Fig.).

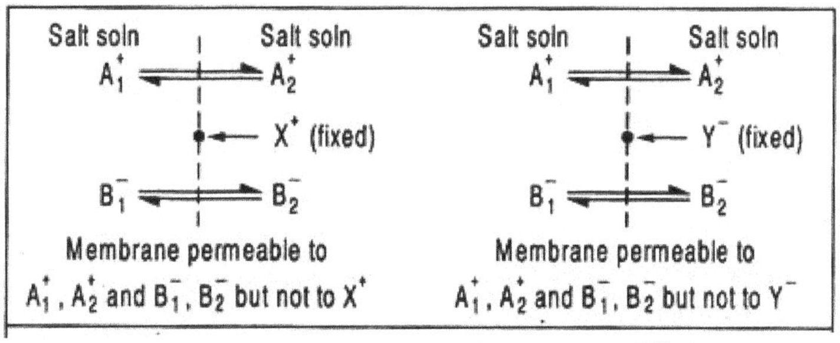

Diagrammatic representation of Donnan equilibrium.

In both the cases represented in the figure X+ and Y– are fixed ions and cannot move from right to left. Due to the presence of these fixed ions some extra ions are absorbed against the concentration gradient. In the first case where X+ is fixed, equal numbers of cations and anions from the left-hand side will diffuse across the membrane until equilibrium is established. This equilibrium would also be electrically balanced.

However, additional anions are needed to balance the positive charges of the fixed cations on the right-hand side of the membrane. Therefore, the anion concentration would become greater on the right-hand side than it is on the left-hand side. Similarly, in the second case where Y– is fixed, cation accumulation takes place at equilibrium.

Thus, the accumulation of ions against a concentration gradient can occur without the participation of metabolic energy until a Donnan equilibrium is reached.

Since the concentrations of mobile ions are unequal on the two sides of the membrane in a Donnan system at equilibrium while the electrochemical potentials are unequal, it follows that there is an electrical potential difference between the two sides. This is sometimes called their Donnan membrane potential.

(d) Mass Flow of Ions:

Many workers believe that ions are absorbed by the roots along with the mass flow of water influenced by the transpiration stream. The theory states that, an increased transpiration rate causes an increase in absorption of ions. Lopushinsky (1964) working with de-topped tomato plants indirectly supported the concept that an increase in transpiration could increase the absorption of salts.

Some workers claim that transpiration pull indirectly affects ion absorption by continuous removal of ions after they have been released into the xylem ducts. Whatever may be the fact, mass flow mechanism may occur in the absence of metabolic energy.

Mediated Uptake

All the experimental findings suggest the concept of mediated transport which states that the transport is accelerated due to the presence of carrier substances in the membrane, which interact with the transported ions or molecules. Ions form a complex with the carrier on the outer side of the membrane. This complex is broken down on the inner side.

The following reaction takes place inside the membrane

C + S = CS where S = Substance and C = Carrier.

The concept has been supported by the (1) radioactive ion exchange, (2) saturation effect, and (3) specificity.

(a) Radioactive Ion Exchange:

Using radioactive ions, Legget and Epstein (1956) observed that the ions failed to diffuse freely through the cell membrane. From this observation they pointed out that the movement of ions across the impermeable membrane might be accomplished by the intervention of carriers.

(b) Saturation Effect:

It has been observed that with the increase in salt concentration of the medium, the rates of ion absorption do not increase beyond limit. That is, a saturation point is reached. This is very analogous with the saturation effect found in enzyme catalyzed reactions, suggesting that all active sites of the carriers are occupied by the ions.

At this point, ion transport is kept constant and cannot be made to proceed faster by an increase in salt concentration. So, the phenomenon of saturation effect indicates the presence of carriers.

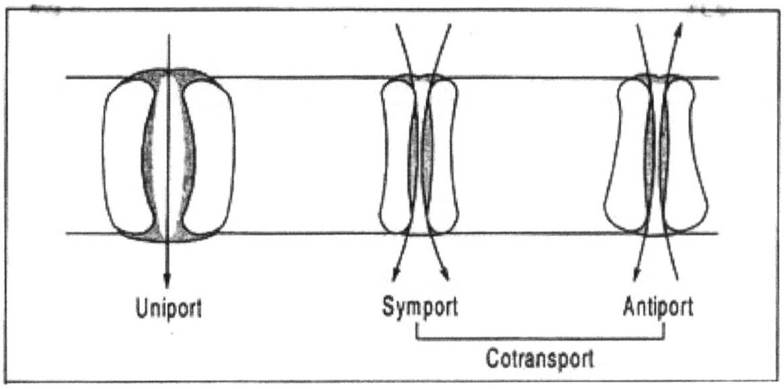

Uniport, symport and antiport translocation systems found in membranes.

(c) Specificity:

Roots absorb ions selectively. Different ions are absorbed at different rates and show different levels of accumulation in the root tissue. This situation is analogous to enzyme-substrate activity. Epstein and Hagen (1952) have observed that the monovalent cations potassium, cesium and rubidium compete with each other for the same binding site.

There are different carriers for different cations and anions. They are also called transporters or perm-eases. They may be of different types such as uniport, symport and antiport. Transporters that carry only one substrate are called uniport systems.

On the other hand, the transporters may carry two ions or solutes simultaneously across a membrane, which are called co-transport systems. When two substrates move in opposite directions, the process is anti-port transport whereas, in symport system, two substrates are moved simultaneously in the same direction.

Mediated transport is classified into two categories depending on the thermodynamics of the system:

1. Passive-mediated transport, or facilitated diffusion, in which a specific molecule flows from high concentration to low concentration.

2. Active transport, in which a specific molecule is transported from low concentration to high concentration, that is, against its concentration gradient. Such an endergonic (energy requiring) process must be coupled to a sufficiently exergonic (energy generating) process to make it favorable (i.e., $\Delta G < 0$).

(d) Passive Mediated Transport:

Ionophores Facilitate Ion Diffusion

One type of carrier molecule is an ionophore, an organic molecule.

Carrier ionophores increase the permeability of membranes to a particular ion by binding the ion, diffusing through the membrane, and releasing it on the other side. For net transport to occur, the uncomplexed ionophore must then return to the original side of the membrane ready to repeat the process (Fig.). The ionic complexes of all carriers must therefore be soluble in nonpolar solvents.

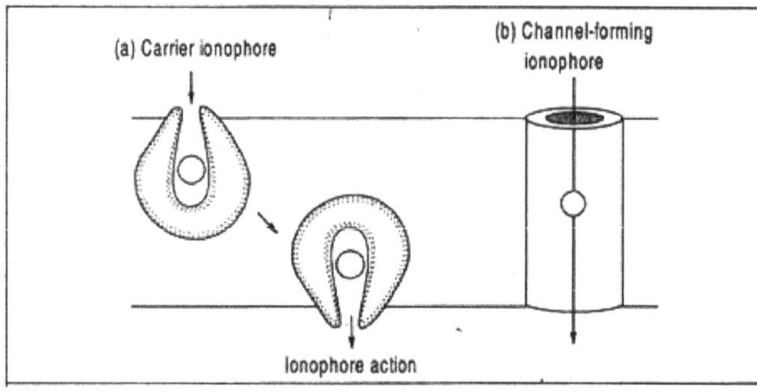

Carrier ionophores transport ions by diffusing through membrane lipids and channel-forming ionophores span the membrane with a channel through which ions can move.

Another type of ionophore called channel-forming ionophores, form solvent-filled, trans membrane channels or pores through which their selected ions can diffuse. Even small amounts of an ionophore greatly increase the permeability of a membrane toward a specific ion.

For example, a single molecule of carrier antibiotic valinomycin transports up to 104K+ ions per second across a membrane. Channel formers, such as the antibiotic gramicidin A, have an even greater ion throughput, over 107K+ ions per second.

Four fundamental classes of transport system are present at all membranes. These are — (1) carriers, (2) pumps, (3) ion channels arid (4) aquaporin's.

Carriers

Unlike pumps, carriers do not catalyse scalar reactions such as ATP hydrolysis.

In other words, the transport process does not involve chemical modification of any of the compounds bound to the carrier. The carriers catalyse only vectorial reactions, i.e., the movement of inorganic ions and simple organic solutes across membranes. The carriers exhibit Michaelis-Menten kinetics that indicates conformational changes during transport.

The array of ions and solutes trans-located by carriers is vast. The principal inorganic nutrients, including NH_4^+, NO_3–, Pi, K^+, and SO_4^{2-}, CI^- are all trans-located into cells by plasma membrane carriers.

The organic solutes like sugars, amino acids, and purine and pyrimidine bases are also trans-located into cells by carriers. The plasma membrane carriers are not only important for nutrient absorption from the soil but also play fundamental roles in the mobilization and storage of metabolites.

For Example

1. The sucrose carriers those are specifically responsible for loading of sucrose into the phloem.
2. Many species reduce NO_3^- in the leaves, and the resulting organic nitrogen compounds are loaded through the carriers into the phloem for transport to sink tissues.

3. During germination, hydrolysis of storage polymers yields sugars and amino acids that need carrier-facilitated mobilization to the growing embryo.

4. Tonoplast carriers catalyse sequestration (temporary storing) of Na^+, Ca^{2+}, Mg^{2+}, and NO_3- as well as sucrose and amino acids into the vacuoles.

5. The triose phosphate trans-locator in the chloroplast membrane exchanges equal amounts of triose phosphate (stromal) for Pi (cytosolic).

6. A carrier in the inner mitochondrial membrane executes the stoichiometric exchange of matrix ATP for cytosolic ADP, maintaining the substrate-product ratio at values favorable for mitochondrial ATP synthesis.

Like the enzymes the carriers are specific for their substrates. Even they are stereo-specific, i.e., specific for D- and L-forms. The acidic, basic and neutral amino acids are transported by distinct plasma membrane carriers. Thus a wide range of carrier types are found in all membranes.

Most of these plant carriers are energized by coupling to the proton motive force (pmf). They behave as H+ – symporters or –anti-porters. Molecular identification of carriers defines them as members of the major facilitator super family. The carriers can be identified genetically.

The most widely applicable and successful approach applied to plant carriers has been yeast complementation. Yeast transport mutants that are defective in growth on the solute of interest are transformed with cDNA from a plant library.

Transformed yeast colonies that are able to grow on the particular solute must accordingly contain functional copies of the relevant plant transport system, which can then be sequenced from the vector. In this way plasma membrane carrier systems for sugars, amino acids, inorganic cations (e.g., K^+), and inorganic anions (e.g., SO_4^{2-}) have been identified in plants.

All the plant carriers identified so far are hydrophobic and contain only single subunit with molecular masses 40 – 50 kDa. The structure of the carrier shows 12 trans membrane spans, with the most extensive hydrophilic loop appearing between trans membrane spans 6 and 7.

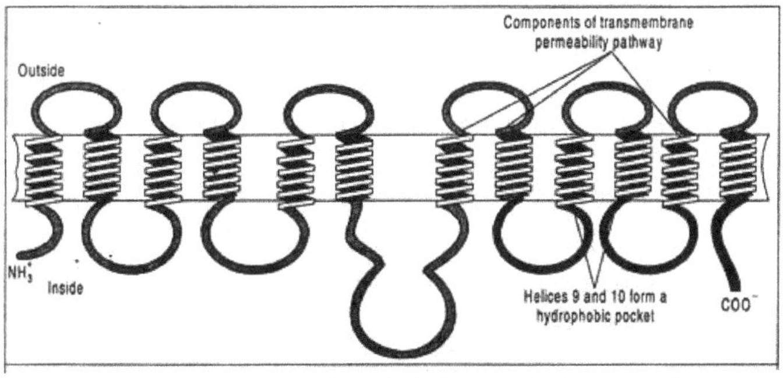

A generalized structural model of a carrier in a membrane.

The motifs present in the proteins are separated by the large hydrophilic loops, indicating that an ancient progenitor of these carrier systems underwent a gene duplication and fusion. Sequence analysis and the general membrane distribution; suggest to place these plant plasma membrane carriers into a large and diverse group of transport systems known as the major facilitator superfamily (MFS).

This group includes both H^+ -coupled sugar transport systems from bacteria and uniporters, carriers from animal cells. Expression of such carriers in particular cell type's gives clues to cell function.

Ion Channels

Ion channels are ubiquitous in plant membranes and are studied by electrophysiological techniques. Ion fluxes through channels are driven solely by electrochemical potential differences. Ion flow through channels is passive. Thus, in contrast to pumps or ion-coupled carrier activity, the direction of flow of a particular ion through a channel is dictated by the electrochemical potential gradient for that ion.

The ion channels exhibit ionic substrate selectivity. Most classes of ion channels in plants are of two types — (1) cation channels and (2) anion channels. Cation channels are further subdivided into K^+ -selective and Ca^{2+}-selective channels. Most plasma membrane anion channels allow a wide range of anions, including CI^-, NO_3^-, and organic acids. Other anion channels in the tonoplast select specifically for malate.

Ion fluxes through channels are monitored as electrical currents. The selectivity of ion channels is due to the specific binding sites located within the channel pores. In some cases, these ion binding sites have been identified at a molecular level.

Ion channels are gated, often by voltage or ligands, through changes in open state probability. The channels are tightly controlled by conformational shifts between permeable (open, O) and non-permeable (closed, C) states.

This alteration between O and C states, known as gating, is represented by the following reaction:

$$\text{Channel}_c \leftrightarrow \text{Channel}_Q$$

In all types of channels gating is controlled by membrane voltage, a ligand, or both. When a gating factor activates a channel, the equilibrium of the above reaction shifts from left to right. Voltage-dependent K^+ channels at the plasma membrane allow controlled K^+ uptake and loss. Time-dependent inward and outward currents are carried by separate classes of ion channel.

The channels carrying these currents are called rectifiers. Like valves, rectifying channels carry current in one direction but not the other. For these reasons, the channels are known as K^+ inward rectifiers and K^+ outward rectifiers.

Inward rectifiers take up K^+ ions not only from the soil but also from the apoplast surrounding most cells. The resulting K^+ accumulation contributes to cell turgor. Inward rectifiers have been cloned by complementation of yeast mutants defective in K^+ uptake.

Plant inward rectifier subunits are products of a multi-gene family, members of which show tissue-specific expression. One member, KAT 1, is expressed selectively in guard cells, whereas another AKT1 is expressed in roots and hydathodes.

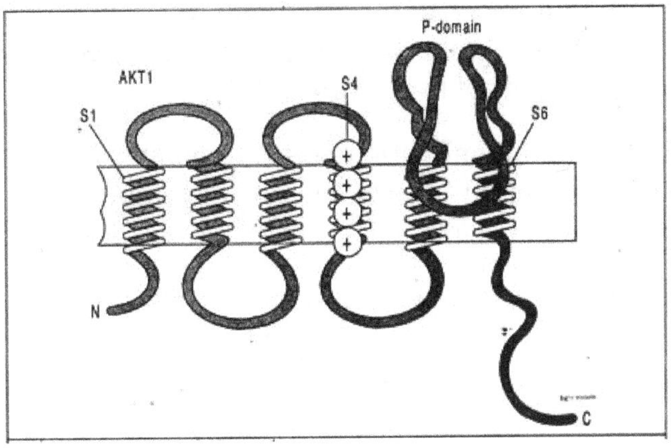

Structure of inward rectifying K* channel AKT1. S1.S6, there are six trans membrane apans with a membrane-intrusive loop between S5 and S6, forming a pore domain (P-domain) of the channel.

Several structural features define plant inward rectifying channels as members of the Shaker family, a superfamily of voltage-gated K^+, Ca^{2+}, and Na^+ channels.

The fourth trans membrane helix (S4 domain) of AKT 1 exhibits a regular pattern of positively charged lysine or arginine residues that tend to project from one side of the helix. This region of the protein forms the voltage sensor, which is involved in opening the channel in response to permissive voltage.

Due to hyperpolarization of the membrane, the S4 helix is thought to screw out of the membrane slightly. This conformational transition opens the "gate" that controls ion flow through a separate part of the protein.

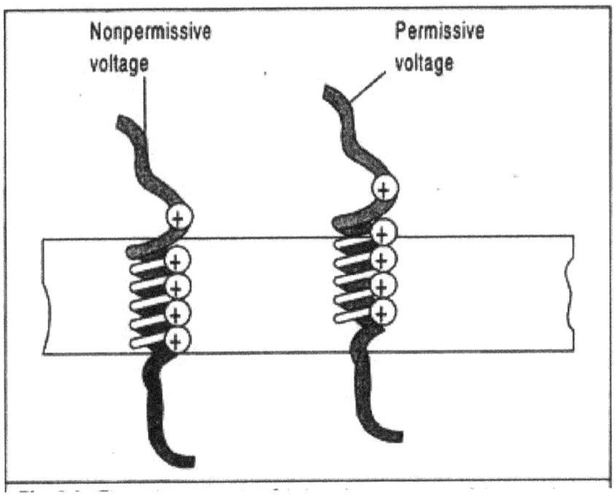

The voltage sensing S4 domain screws out of the membrane in response to application of an activating voltage.

The voltage-insensitive cation channels are major pathways for Na^+ uptake across the plasma membrane and for salt release into the xylem. Such channels give plants tolerance to salinity. These channels are partially blocked by external Ca^{2+}.

Active Uptake of Ion Absorption (Metabolic)

Active uptake of ions is one of the most important features of life processes. It is accomplished through the coupling of diffusion fluxes to the exergonic reactions that take place in the bulk of the membrane. The transfer of ions occurs at the expense of the free energy liberated in chemical reactions. As a rule, this is the energy of hydrolysis of ATP.

It has been observed that both anions and cations are accumulated by plants against concentration gradients to a great extent that cannot be explained by any known electrochemical mechanism, such as ion exchange, Donnan equilibrium, etc.

It has also been observed that the rate and amount of absorption of salts is directly related to the expenditure of metabolic energy. Hober (1945), reported that fresh water alga Nitella, absorbed potassium ions to a concentration 1000 times greater than the concentration of K^+ ions in the surrounding medium.

This type of absorption where the concentration of ions is much higher within the cells than in the external solution is called accumulation.

The extent to which the concentration becomes greater internally than externally is called the accumulation ratio. Thermodynamic laws show that free diffusion and other passive mechanisms not involving expenditure of metabolic energy could not be responsible for such great accumulation.

Many studies indicate that solute transport into cells is strongly dependent upon metabolic energy. Ion accumulation is inhibited when the metabolic activity of the plant is inhibited by low temperatures, low oxygen tension, metabolic inhibitors, and so on. Many workers like Steward (1932), Hopkins (1956) and others have found a parallel curve for salt accumulation and respiration as a function of oxygen tension.

With the decrease in oxygen content of the medium, ion accumulation decreases and ultimately stops completely. Robertson and Turner (1945) and Lundegardh (1955) have reported that metabolic inhibitors like azides and cyanides inhibit ion accumulation with the inhibition of respiration.

An additional evidence in support of the existence of an active process of ion uptake is provided by the phenomenon of salt induced respiration. Lundegardh and Burstrom (1933) observed that the rate of respiration increases when a plant or tissue is transferred from water to a salt solution. The amount by which respiration is increased over the normal rate has been called salt respiration or anion respiration.

Mechanism of Active Uptake

Now, it is necessary to evaluate the concentration effect and the electro potential effect to understand whether the ion absorption takes place under the influence of metabolic energy.

The electrochemical potential difference across the membrane is designated as $\Delta\mu$ and expressed by the following modified Nernst equation:

$\Delta\mu = \Delta\,(RT\ \text{In}C) + \Delta\,(zFE)$ ['z' is the algebraic valency]

where $\Delta\,(RT\ \text{In}C)$ = chemical potential difference due to concentration effect

$\Delta\,(zFE)$ = electro potential difference

R = universal gas constant

T = absolute temperature

C = ion concentration inside and outside the membrane

F = Faraday's constant

E = electro potential in volts on each side of the membrane

The Above Equation Can be Simplified as Follows:

$\text{Log } Ci/Co = -zF\Delta E/2.3RT$

where Ci = inside ion concentration

Co = outside ion concentration

Ci/Co is the predicated ratio at equilibrium, when $\mu = 0$. If the actual-ratio is greater than the predicted ratio, the cell is performing work to absorb ions using energy. This is called active absorption. Passive absorption occurs when the actual ratio is equal to or lesser than the predicated value.

The active ion transport may be (1) primary and (2) secondary.

Primary Active Transport

The primary active transport is coupled directly to a source of energy other than electrochemical potential gradient, such as ATP hydrolysis, an oxidation reduction reaction, etc. The membrane proteins that carry out primary active transport are called pumps. Most pumps transport ions. Ion pumps are further characterized as either electro genic or electro neutral.

In general, electro genic transport refers to ion transport involving the net movement of charge across the membrane. In contrast, electro neutral transport, as the name implies, involve no net movement of charge. For example, the Na^+/K^+ -ATPase of animal cells pumps three Na^+ ions out of every two K^+ ions in, resulting in a net outward movement of one positive charge.

Secondary Active Transport

Secondary active transport uses the energy stored in electrochemical-potential gradient. Protons are extruded from the cytosol by electro genic H^+ ATPase operating in the plasma membrane and tonoplast. Consequently, a membrane potential and a pH gradient are created at the expense of ATP hydrolysis.

This gradient of electrochemical potential for H^+, the proton motive force represents stored free energy in the form of H^+ gradient.

Carbon Cycle in Soil

Carbon (C) is one of the most common elements in the universe and found virtually everywhere on earth: in the air, the oceans, soil, and rock. Carbon is part of geologic history in rock and especially the ancient deposits that formed coal, oil and other energy sources we use today. Carbon is also an essential building block of life and a component of all plants and animals on the planet. It has unique bonding properties that allow it to combine with many other elements. These properties enable the formation of molecules that are useful and necessary to support life. The role of carbon in living systems is so significant that a whole branch of study is devoted to it: organic chemistry. Carbon that is not tied up in rock or deep in the oceans is constantly changing and moving. This process is called the carbon cycle. Soil holds the largest portion of active carbon on earth. Plants take carbon from the air and convert it to plant tissue, some of which returns to the soil as plant residue.

Agriculture's Role in the Carbon Cycle

Carbon is critical to soil function and productivity, and a main component of and contributor to healthy soil conditions. Soil management plays a critical role in whether the carbon remains in the soil or is released to the atmosphere. Agricultural practices can impact both the amount and the composition of soil organic carbon and hence also the soil's physical, biological, and chemical condition, the combination of things that defines soil health. Farm practices that affect carbon therefore impact agricultural productivity and resilience (the soil's ability to deal with weather extremes) and the carbon cycle itself.

Importance of Soil Organic Carbon

While the agricultural sector has the ability to impact the carbon cycle on a large scale, often through the release of carbon, farmers have a vested interest in retaining and increasing soil organic carbon for individual fields because soil and yield tend to improve when the soil organic carbon level increases. Higher soil organic carbon promotes soil structure or tilth meaning there is greater physical stability. This improves soil aeration (oxygen in the soil) and water drainage and retention, and reduces the risk of erosion and nutrient leaching. Soil organic carbon is also important to chemical composition and biological productivity, including fertility and nutrient holding capacity of a field. As carbon stores in the soil increase, carbon is "sequestered", and the risk of loss of other nutrients through erosion and leaching is reduced. An increase in soil organic carbon typically results in a more stable carbon cycle and enhanced overall agricultural productivity, while physical disturbances of the soil can lead to a net loss of carbon into the surrounding environment due to formation of carbon dioxide (CO_2).

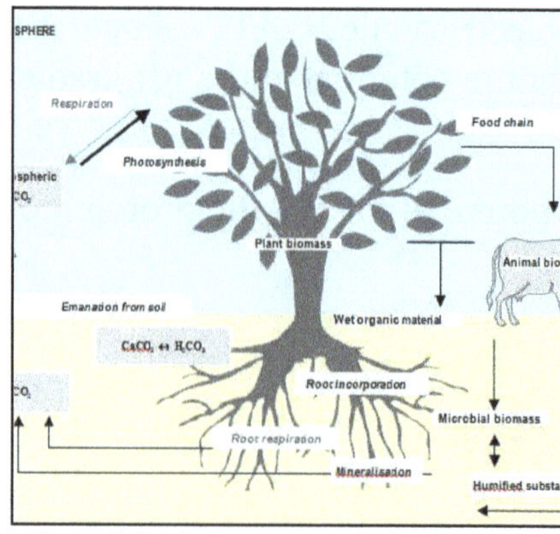

The carbon cycle and the interaction of plant, soil, and atmosphere.

Management Practices for C Sequestration

With agricultural productivity so dependent on soil organic carbon and carbon cycling, how can we best manage fields to enhance soil organic carbon levels while also reducing carbon loss into the atmosphere?

Table. Management practices that can increase soil organic carbon and reduce carbon loss into the atmosphere.

Management practices	Functions and explanation
Conservation tillage practices	Conservation tillage practices including no-till management aid in storing soil organic carbon, keeping the physical stability of the soil intact. When reduced-till systems are combined with residue management and manure management, soil organic carbon can increase over time.
Crop residue management	Returning crop residue to the soil adds carbon and helps to maintain soil organic matter.
Cover crops	Cover crops can increase soil carbon pools by adding both root and above ground biomass. Covers also reduce the risk of soil erosion and the resulting loss of carbon with soil particles. Cover crops also enhance nutrient cycling and increase soil health over time.
Manure and compost	Adding organic amendments such as manure or compost can directly increase soil carbon, and also result in increased soil aggregate stability. This enhances the biological buffering capacity of the soil, resulting in greater yields and yield stability over time.
Crop selection	Perennial crops eliminate the need for yearly planting and increase soil organic carbon by root and litter decomposition post-harvest. Crops with greater root mass in general add to root decomposition and physically bond aggregates together. Using high residue annual crops can also help reduce net carbon loss from cropping systems.

The ability of agricultural fields to sequester carbon (capture and storage of carbon that would otherwise be lost to the environment) depends on several factors including climate, soil type, type of crop or vegetation cover, and management practices. Employing farming practices that reduce disturbance of the soil (less aeration from tillage helps protect carbon), combined with practices that bring additional carbon to the soil, will allow for carbon sequestration over time. Such practices include implementation of conservation tillage (no-till, zone-till, minimum-till, shallow mixing or injection for manure applications), retaining crop residues, including cover crops in crop rotations, adding organic nutrient sources such as manure and compost, and including perennial crops in crop rotations (Table). Their implementation may slow or even reverse the loss of carbon from agricultural fields, improve nutrient cycling and reduce nutrient loss.

Soil carbon management is an important strategy for improving soil quality, increasing crop yields, and reducing soil loss. Capturing carbon in the soil helps improve soil health and productivity, and stabilize the global carbon cycle, benefiting agricultural production.

Nitrogen Cycle in Soil

Nitrogen exists in soils in many forms and constantly changes from one form to another. The paths that the different forms of nitrogen follow through the ecosystem are collectively called the nitrogen cycle. Understanding how the different pools of nitrogen interact and the processes by which these forms enter and leave the cycle is the subject of continuing study. Nitrogen is found in both inorganic and organic forms. Soil organic matter is composed primarily of amides (NH_2) and accounts for more than 90 percent of the total nitrogen present in most environments. In general, nitrogen is not found associated with soil minerals, as is the case with phosphorus. Some clay minerals may tie up small amounts of nitrogen in the ammonium form (clay fixation) but not in the same magnitude as phosphorus.

Organic Nitrogen

Soil organic matter is the major storehouse of many plant nutrients in soils, including nitrogen, phosphorus, sulfur, calcium and magnesium. Soil organic matter is composed of a stable material called humus, an easily decomposed material (litter), soil microbes and some other organic molecules. Typically, humus will contain 45 to 55 percent carbon and about 5 percent nitrogen. In other words, soil humus typically has a carbon to nitrogen ratio (C:N ratio) of approximately 12:1 for surface soils and 8-10:1 for subsurface soils.

Mineralization

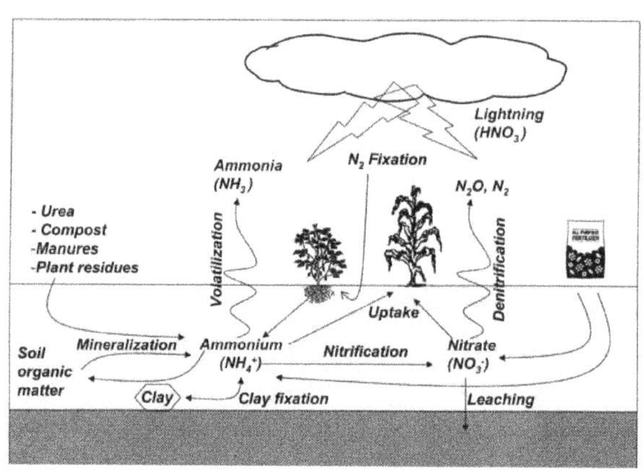

Nitrogen cycle in upland soils.

Mineralization

Organic N ⟶ Ammonia (NH_3^+) ⟶ Ammonium (NH_4^+)

Nitrification

Ammonium (NH_4^+) ⟶ Nitrate (NO_3^-)

Immobilization

Organic N ⟵ (NH_4^+) (NO_3^-)

Nitrogen mineralization/immobilization process.

Nitrogen that is present in soil organic matter, crop residues and manures is converted to the inorganic form by the process of mineralization. Initially, larger organic matter molecules are broken down into smaller ones, with soil microorganisms attacking these remaining materials by producing specific enzymes. The transformation of organic nitrogen to the ammonia (NH_3) and ammonium ($NH4^+$) forms is referred to as ammonification. The resultant ammonium can then be transformed to the nitrate form (NO_3^-) by a process called nitrification. Since the decomposition process is carried out by living organisms, it is affected by several environmental variables, including soil moisture, temperature, pH, the C:N ratio and the type of organic materials in the residue. Cotton and grain sorghum stalks and small grain residues are relatively high in carbon and low in nitrogen (having C:N ratios greater than 30:1). Soil organisms may need additional nitrogen to decompose these residues. When the nitrogen supply is limited, soil microbes compete with plants for fertilizer N in a process called immobilization. When these microbes die, nitrogen tied up in the decomposition process again becomes available for crop use.

Biological Nitrogen Fixation

Atmospheric nitrogen (N_2) is basically an endless source of N, but this nitrogen cannot be used directly by most plants. Legumes such as alfalfa, soybeans and clovers form a symbiotic association (mutually beneficial) with specific bacteria to convert atmospheric N_2 to a form plants can use. The plant provides nutrients and other compounds to the N_2-fixing bacteria, and in return, the plant benefits from the N fixed by the microorganisms. The amount of N fixed varies among plants and growing conditions (Table). The symbiotic association is highly specific; thus, the bacterial species that fixes N2 with soybeans is not effective for fixing N_2 with alfalfa. The site of the N_2-fixing process is a root nodule that forms on the root system and has a pink color if actively fixing N_2.

The biological nitrogen fixation process is catalyzed (accelerated) by the nitrogenase enzyme, which is affected by a number of soil and weather factors. The association does not work well in soils with very low pH or when high levels of available mineral nitrogen are present. The absence of the right type of inoculum (bacteria) in the soil also will limit nodulation and N_2 fixation.

Table: Typical Amount of Nitrogen Fixed by Selected Legumes

Crop	Amount fixed (lb/A/year)
Alfalfa	170-270
Soybeans	70-170
Clovers	120-170
Vetch	70-150

Some other soil organisms are capable of fixing N2 via non-symbiotic associations. This process is of little significance in most agricultural systems, with the exception of blue-green algae which live in rice systems. Some nitrogen also is fixed by lightning, but in considerably smaller amounts than biologically fixed N_2.

Leaching

Nitrogen in the nitrate form is very mobile and highly soluble in water. Rainfall moving through the root zone may wash nitrate downward, reaching tiles or drainage channels and potentially reaching groundwater or surface waters. Leaching is a more serious problem in highly permeable sandy soils than in clayey soils. The magnitude of nitrate loss through leaching depends on the amount and intensity of the rainfall or irrigation water and the amount of nitrate present in the soil.

Loss of nitrate by leaching is of concern for two important reasons. Nitrate below the root zone is no longer available for plant use, representing an important loss of resources. Water quality problems caused by excess nitrogen leaving fields may result in the deterioration of drinking water sources and wildlife habitat.

Volatilization as NH3

A significant amount of nitrogen fertilizer can be lost through the process of volatilization if not properly applied. Many commercial fertilizers contain ammonium or convert readily to ammonium, such as urea. Under alkaline pH conditions (pH > 7), a percentage of the ammonium (NH_4^+) can be converted to ammonia gas (NH_3) and escape to the atmosphere. In addition to soil pH, high soil temperature, excessive soil moisture and strong winds will contribute to such loss.

Ammonia volatilization also can occur under neutral and acidic soil pH conditions. During the transformation of urea to ammonium (NH_4^+), the pH of soil adjacent to the fertilizer can increase to levels that promote ammonia formation. This effect is more pronounced in sandy and silt loam soils, which lack the ability to resist pH changes (i.e., low pH buffering capacity).

Denitrification

The process of denitrification is a form of respiration carried out by microorganisms under low oxygen conditions. Specific microorganisms, primarily bacteria, have the ability to use nitrate instead of oxygen to carry out their metabolic functions. In this process, nitrate (NO3-) is reduced to NO2- and then to gaseous forms including N2O and N2.

Temperatures between 75 and 95 degrees F, low soil oxygen levels for several days (waterlogged conditions), soil pH > 6 and the presence of a carbon source (organic matter) are ideal conditions for denitrification to proceed at a high rate. Denitrification varies widely among locations and time of year, but it normally represents only a small percentage of total N loss. The process of denitrification is detrimental, since plant-available nitrogen is lost to the atmosphere. However, denitrification also can be used as a means to prevent the buildup of nitrates in groundwater wells, manure storage lagoons and wastewater treatment plants.

Nitrogen Cycle in Flooded Soils

In upland soils, the process of mineralization gradually converts organic nitrogen to ammonium and then to nitrate. Under flooded conditions, due to the low oxygen concentration, the end product of the mineralization process is ammonium (Figure).

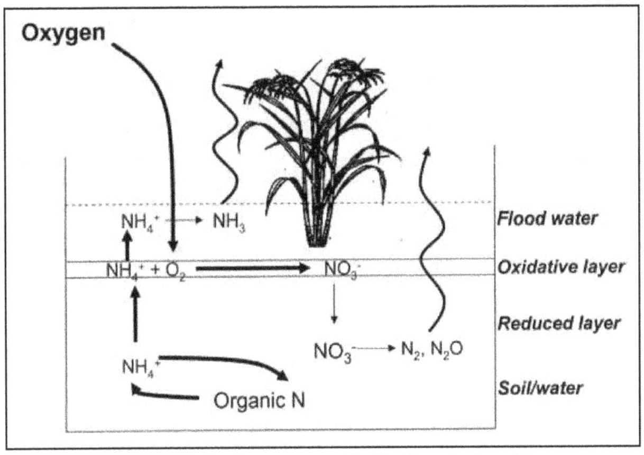

Simplified nitrogen cycle in flooded soils.

In flooded soils, such as rice paddies, two layers are formed – an oxidative layer at the soil/water interface, which is less than one inch thick, and a reductive layer beginning immediately beneath. If nitrate fertilizer is added to flooded soil, bacteria can convert it to nitrogen gas (denitrification), which may be lost to the atmosphere. Ammonium fertilizer placed in the oxidative layer also may be converted to nitrates, just as in an upland soil.

Phosphorous Cycle in Soil

Phosphorus, Crops and the Environment

This agronomy fact sheet provides a brief overview of the important components of the phosphorus (P) cycle. Understanding the P cycle can help producers make decisions regarding P management on the farm, both for farm profitability and protection of the environment.

Most plants are only about 0.2% P by weight, but that small amount is critically important. Phosphorus is an essential component of adenosine triphosphate (ATP), which is involved in most biochemical processes in plants and enables them to extract nutrients from the soil. Phosphorus also plays a critical role in cell development and DNA formation. Insufficient soil P can result in delayed crop maturity, reduced flower development, low seed quality, and decreased crop yield. Too much P, on the other hand, can be harmful in some situations; when P levels increase in fresh water streams and lakes, algae blooms can occur. When algae die, their decomposition results in oxygen depletion which can lead to the death of aquatic plants and animals. This process is called "eutrophication".

Crop Uptake

One goal with field crop management is to optimize crop uptake of available P. A typical corn silage crop will remove about 4.3 lbs of P2O5 per ton of silage (35% dry matter). Soil testing of available P can help avoid application of fertilizer P that is not needed for optimum production. Applying fertilizer beyond crop needs is a waste of time and money, and can be harmful to the environment.

Phosphorus Cycle

Phosphorus exists in many different forms in soil. For practical purposes, we can group these sources into four general forms: (1) plant available inorganic P, and three forms which are not plant available: (2) organic P, (3) adsorbed P, and (4) primary mineral P. The P cycle in Figure shows these P forms and the pathways by which P may be taken up by plants or leave the site as P runoff or leaching. The general P transformation processes are: weathering and precipitation, mineralization and immobilization, and adsorption and desorption. Weathering, mineralization and desorption increase plant available P. Immobilization, precipitation and adsorption decrease plant available P.

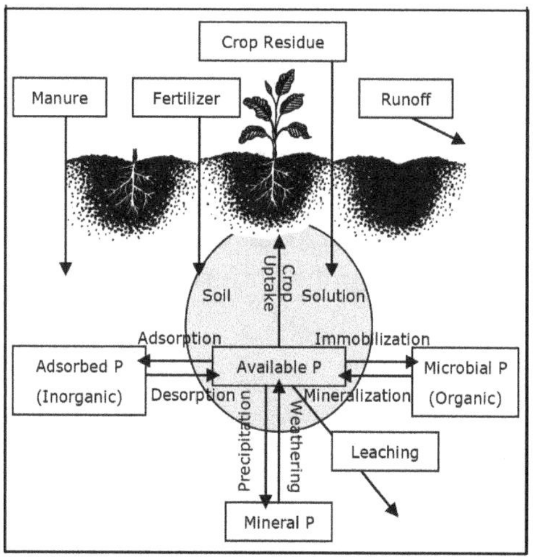

Simplified phosphorus cycle.

Weathering and Precipitation

Soils naturally contain P-rich minerals, which are weathered over long periods of time and slowly made available to plants. Phosphorus can become unavailable through precipitation, which happens if plant available inorganic P reacts with dissolved iron, aluminum, manganese (in acid soils), or calcium (in alkaline soils) to form phosphate minerals.

Mineralization and Immobilization

Mineralization is the microbial conversion of organic P to $H_2PO_4^-$ or HPO_4^{2-}, forms of plant available P known as orthophosphates.

Immobilization occurs when these plant available P forms are consumed by microbes, turning the P into organic P forms that are not available to plants. The microbial P will become available over time as the microbes die.

- Maintaining soil organic matter levels is important in P management. Mineralization of organic matter results in the slow release of P to the soil solution during the growing season, making it available for plant uptake. This process reduces the need for fertilizer applications and the risk of runoff and leaching that may result from additional P.
- Soil temperatures between 65 and 105°F favor P mineralization.

Adsorption and Desorption

Adsorption is the chemical binding of plant available P to soil particles, which makes it unavailable to plants. Desorption is the release of adsorbed P from its bound state into the soil solution.

- Adsorption (or "fixing" as it is sometimes called) occurs quickly whereas desorption is usually a slow process.
- Adsorption differs from precipitation: adsorption is reversible chemical binding of P to soil particles while precipitation involves a more permanent change in the chemical properties of the P as it is removed from the soil solution.
- Soils that have higher iron and/or aluminum contents have the potential to adsorb more P than other soils.
- Phosphorus is in its most plant available form when the pH is between 6 and 7. At higher pH, P can precipitate with Ca. At lower pH, P tends to be sorbed to Fe and Al compounds in the soil.
- Every soil has a maximum amount of P that it can adsorb. Phosphorus losses to the environment through runoff and/or leaching increase with P saturation level.
- Precise fertilizer placement can decrease P adsorption effects by minimizing P contact with soil and concentrating P into a smaller area. Band application of fertilizer is a common example of this.

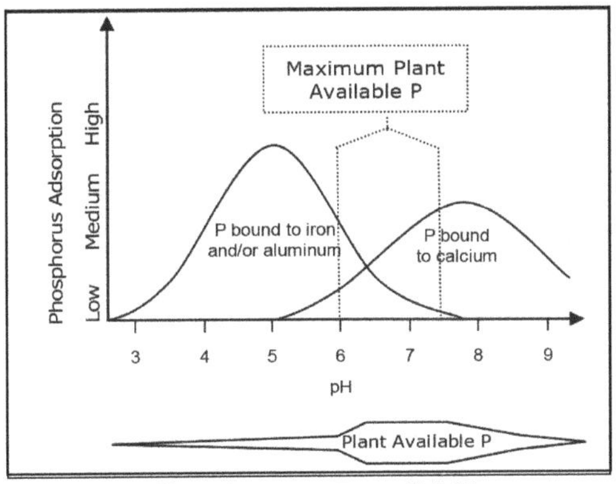

Soil pH impacts P availability.

Runoff

Runoff is a major cause of P loss from farms. Water carries away particulate (soil-bound) P in eroded sediment, as well as dissolved P from applied manure and fertilizers. Erosion control practices decrease P losses by slowing water flow over the soil surface and increasing infiltration.

Leaching

Leaching is the removal of dissolved P from soil by vertical water movement. Leaching is a concern in relatively high P soils (near or at P saturation), especially where preferential flow or direct connections with tile drains exist.

Crop uptake is the goal of applying P fertilizer or manure to the soil. If soil tests P levels are already optimum, P additions through fertilizer or manure should not exceed crop removal. If additional P is needed (soils testing low or medium in P), P adsorption can be minimized by band applications and by maintaining an optimum pH. Naturally occurring immobilization of P by microbes can help ration plant available P to crops over the course of a growing season. Steps should be taken to reduce losses in order to maximize the efficiency of fertilizer and manure applications.

Potassium Cycle in Soil

Potassium: Nature and Ecological Roles: Potassium is an essential plant nutrient that sometimes limits plant growth. Potassium is only present in soil as a positively charged cation ($K+$). Its entire lifecycle in the soil is actually linked mostly with cation exchange and mineral weathering. Potassium doesn't cause any nutrient pollution problems and is generally pretty inert.

Potassium in Plant and Animal Nutrition: Potassiums main function in plants and animals is actually not to be synthesized into organic compounds, rather, it activates enzymes. It is the activator for over 80 enzymes responsible for living functions. Potassium always remains in the ionic form as well. Potassium is linked to plants having a really good response to environmental stress as well. It improves a plants hardiness, drought tolerance, and disease resistance. This is an essential nutrient to help plants flower or produce fruit. Potassium is necessary for animals including humans and helps regulate the central nervous system and maintain healthy blood vessels.

Deficiency Symptoms in Plants: Potassium deficiency in plants creates a foliar yellowing on the tips and edges of the oldest leaves first due to its mobility in plants. Certain plants, mainly legumes, will get while spots towards the tips of their leaves. To tell it apart from salinity damage looks for these symptoms in old vs new growth.

The Potassium Cycle: Potassium comes from primary minerals in soil like micas and potassium feldspar. These minerals slowly weather and eventually the potassium becomes more and more available to be held in soil solution or taken up by plant roots. Once plants take up a ton of potassium it is leached from leaves to soil by rainfall. This and animal urine is how it gets returned to the soil. Some potassium is lost to soil losses, runoff, and leaching to groundwater. When plants are harvested and taken away from the soil entirely, the nutrients including potassium goes to wherever the plant matter goes thus ag. soils need potassium amendments. Soils amendments high in potassium include poultry manure and wood ash.

The Potassium Problem in Soil Fertility

Availability of Potassium: Most mineral soils hold a larger volume of potassium than phosphorus and actually the total quantity of soil potassium is larger than any other major nutrient. However, available potassium is found in small amounts in the soil.

Leaching Losses: Potassium is mostly lost from the soil by leaching however its positive structure is great for binding to negative cation exchange sites on clay or humus. Liming can help keep potassium in the soil due to complementary ion effect (K^+ ions can more easily replace Ca^{2+} ions in cation exchange).

Plant Uptake and Removal: Plants take up about as much potassium as they do nitrogen. The removal mostly occurs when plants are harvested and most of the plant matter is removed from the soil it was grown.

Luxury Consumption: Plants will actually take up more potassium than they need when available quantities are large. This means that anything a plant takes up above its required potassium will be wasted if the plant is removed from where it grew. High potassium levels in plants could actually inhibit calcium and magnesium uptake causing nutritional imbalance.

Forms and Availability of Potassium in Soils: There are four forms of potassium involved in the soil potassium cycle: 1. K in mineral structures (unavailable), 2. K in nonexchangeable mineral forms (slowly available), 3. Exchangeable K on colloid surfaces (readily available), and 4. Water soluble K ions (readily available). The exchange of potassium between these four forms is a function of the types of clay minerals. Soils with 2:1 clays have the most potassium. Some plants can actually obtain potassium from generally unavailable forms on primary mineral structures.

Relatively Unavailable Forms: This includes about 90-98% of all soil potassium. This is the potassium held up in feldspar or mica. This form can release potassium slowly over the course of many years. This weathering of the primary potassium minerals is generally assisted by organic/inorganic acids and acidic clays and humus.

Readily Available Forms: Only about 1-2% of total soil potassium. This form of K exists in two forms: 1. In the soil solution, 2. Exchangeable on colloid surfaces. Most of this is in the exchangeable form on charged colloid sites. The free K in soil solution is mostly used by higher plants. There is a continued equilibrium in soils which keeps potassium evenly in solution and on exchange sites.

Slowly Available Forms: Sometimes in type 2:1 mineral clay soils, potassium as well as the similarly sized ammonium molecule will get permanently affixed between layers of growing soil colloids. These aren't held on exchange sites and so are "nonexchangeable ions". This however, acts as an important reservoir for slowly released ions.

Release of Fixed Potassium: There is a lot of nonexchangeable/fixed potassium in soils. It is continually released to an exchangeable form in amounts large enough to be important. There is a good equilibrium between exchangeable K and nonexchangeable. This means some soils, especially sandy soils, have lower CEC and thus don't maintain potassium well. More clayey soils have a better time maintaining enough potassium ions throughout a growing season.

Factors Affecting Potassium Fixation in Soils: There are four soil conditions that influence the amount of K that can be fixed: 1. Types of soil colloids, 2. Wet and dry cycles, 3. Freeze and thaw cycles, 4. The presence of lime.

Effects of Type of Clay and Moisture: Type 1:1 clays fix a small amount of potassium while 2:1 clays fix K in large quantities. Freeze/thaw and wet/dry cycles help to stimulate the exchange equilibrium of K.

Influence of pH: Lime application generally increases soil pH and potassium fixation. In acidic soils, colloids are holding H+ ions preventing K+ from getting close to the exchange sites thus keeping K+ from fixating. So as the pH increases, more K is able to be fixed to soil colloids. Another factor is root uptake, the more calcium and magnesium in soils the more competition these cations have for root uptake and thus less potassium could be absorbed.

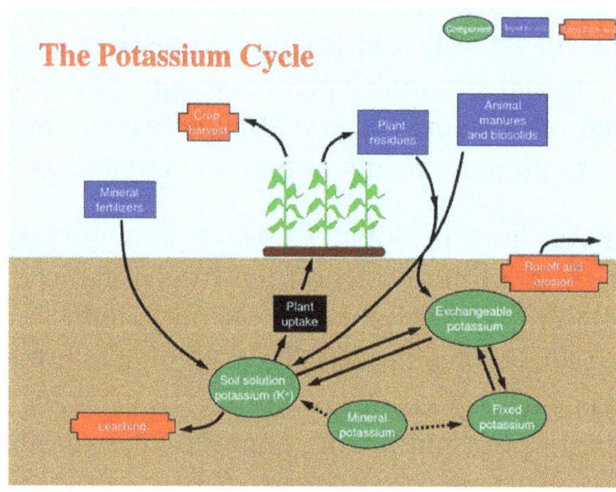

Potassium Cycle.

Practical Aspects of Potassium Management: In most soils, the potassium fertility issue comes with the rate at which potassium can be converted to a plant available form. When plant material isn't removed, cycling between plant matter and soil is usually adequate to supply the next season's plants. A tricky thing about potassium management is that K uptake by plants is not consistent throughout the growing season. Many farmers rely on potassium additions with fertilizer however, excessive K depress Ca and Mg which can cause disruption in plant and animal health.

Frequency of Application: The author suggests that it's best to apply small amounts of potassium to fields every so often in order to prevent excess leaching or over absorption of K by plants. Application would generally increase over many years until only maintenance levels were needed, soils would become built up enough to supply a good amount of potassium by themselves.

5

Pedology: A Study of Soil

Pedology is a scientific discipline which deals with the formation and evolution of soil. In order to completely understand pedology, it is necessary to understand the processes related to it. The following chapter elucidates the varied processes and mechanisms associated with this area of study such as pedogenesis, soil morphology, active layer and soil map.

Pedology is a scientific discipline concerned with all aspects of soils, including their physical and chemical properties, the role of organisms in soil production and in relation to soil character, the description and mapping of soil units, and the origin and formation of soils. Accordingly, pedology embraces several sub disciplines, namely, soil chemistry, soil physics, and soil microbiology. Each employs a sophisticated array of methods and laboratory equipment not unlike that used in studies of the physics, chemistry, or microbiology of non-soils systems. Sampling, description, and mapping of soils is considerably simpler, however. A soil auger is used to obtain core samples in places where no subsurface exposure can be found, and the soil units are defined, delineated, and mapped in a manner similar to procedures in stratigraphy. Such soils studies, in fact, overlap the concerns of the stratigrapher and the geologist, both of whom may treat the soils layers as strata of the Quaternary Period (from 2.6 million years ago to the present).

Soil is not only a support for vegetation, but it is also the zone beneath our feet (the pedosphere) of numerous interactions between climate (water, air, temperature), soil life (micro-organisms, plants, animals) and its residues, the mineral material of the original and added rock, and its position in the landscape. During its formation and genesis, the soil profile slowly deepens and develops characteristic layers, called 'horizons', while a steady state balance is approached.

Soil users (such as agronomists) showed initially little concern in the dynamics of soil. They saw it as medium whose chemical, physical and biological properties were useful for the services of agronomic productivity. On the other hand, pedologists and geologists did not initially focus on the agronomic applications of the soil characteristics (edaphic properties) but upon its relation to the nature and history of landscapes. Today, there's an integration of the two disciplinary approaches as part of landscape and environmental sciences.

Pedologists are now also interested in the practical applications of a good understanding of pedogenesis processes (the evolution and functioning of soils), like interpreting its environmental history and predicting consequences of changes in land use, while agronomists understand that the cultivated soil is a complex medium, often resulting from several thousands of years of evolution. They understand that the current balance is fragile and that only a thorough knowledge of its history makes it possible to ensure its sustainable use.

Key Points

- Complexity in soil genesis is more common than simplicity.
- Soils lie at the interface of Earth's atmosphere, biosphere, hydrosphere and lithosphere. Therefore, a thorough understanding of soils requires some knowledge of meteorology, climatology, ecology, biology, hydrology, geomorphology, geology and many other earth sciences and natural sciences.
- Contemporary soils carry imprints of pedogenic processes that were active in the past, although in many cases these imprints are difficult to observe or quantify. Thus, knowledge of paleoecology, palaeogeography, glacial geology and paleoclimatology is important for the recognition and understanding of soil genesis and constitute a basis for predicting the future soil changes.
- Five major, external factors of formation (climate, organisms, relief, parent material and time), and several smaller, less identifiable ones, drive pedogenic processes and create soil patterns.
- Characteristics of soils and soil landscapes, e.g., the number, sizes, shapes and arrangements of soil bodies, each of which is characterized on the basis of soil horizons, degree of internal homogeneity, slope, aspect, landscape position, age and other properties and relationships, can be observed and measured.
- Distinctive bioclimatic regimes or combinations of pedogenic processes produce distinctive soils. Thus, distinctive, observable morphological features, e.g., illuvial clay accumulation in B horizons, are produced by certain combinations of pedogenic processes operative over varying periods of time.
- Pedogenic (soil-forming) processes act to both create and destroy order (anisotropy) within soils; these processes can proceed simultaneously. The resulting soil profile reflects the balance of these processes, present and past.
- The geological Principle of Uniformitarianism applies to soils, i.e., pedogenic processes active in soils today have been operating for long periods of time, back to the time of appearance of organisms on the land surface. These processes do, however, have varying degrees of expression and intensity over space and time.
- A succession of different soils may have developed, eroded and/or regressed at any particular site, as soil genetic factors and site factors, e.g., vegetation, sedimentation, geomorphology, change.
- There are very few old soils (in a geological sense) because they can be destroyed or buried by

geological events, or modified by shifts in climate by virtue of their vulnerable position at the surface of the earth. Little of the soil continuum dates back beyond the Tertiary period and most soils and land surfaces are no older than the Pleistocene Epoch. However, preserved/lithified soils (paleosols) are an almost ubiquitous feature in terrestrial (land-based) environments throughout most of geologic time. Since they record evidence of ancient climate change, they present immense utility in understanding climate evolution throughout geologic history.

- Knowledge and understanding of the genesis of a soil is important in its classification and mapping.
- Soil classification systems cannot be based entirely on perceptions of genesis, however, because genetic processes are seldom observed and because pedogenic processes change over time.
- Knowledge of soil genesis is imperative and basic to soil use and management. Human influence on, or adjustment to, the factors and processes of soil formation can be best controlled and planned using knowledge about soil genesis.
- Soils are natural clay factories (clay includes clay mineral structures and particles less than 2 μm in diameter). Shales worldwide are, to a considerable extent, simply soil clays that have been formed in the pedosphere and eroded and deposited in the ocean basins, to become lithified at a later date.

Pedogenesis

Pedogenesis can be defined as the process of soil development. Late in the 19th century, scientists Hilgard in the United States and the Russian Dukuchaev both suggested independently that pedogenesis was principally controlled by climate and vegetation. This idea was based on the observation that comparable soils developed in spatially separate areas when their climate and vegetation were similar. In the 1940s, Hans Jenny extended these ideas based on the observations of many subsequent studies examining the processes involved in the formation of soils. Jenny believed that the kinds of soils that develop in a particular area are largely determined by five interrelated factors: climate; living organisms; parent material; topography; and time (Figure).

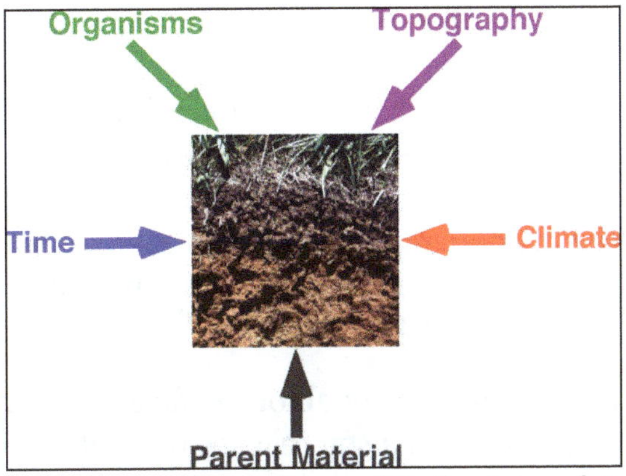

The development of a soil is influenced by five interrelated factors: organisms, topography, time, parent material, and climate.

Climate plays a very important role in the genesis of a soil. On the global scale, there is an obvious correlation between major soil types and the Koppen climatic classification systems major climatic types. At regional and local scales, climate becomes less important in soil formation. Instead, pedogenesis is more influenced by factors like parent material, topography, vegetation, and time. The two most important climatic variables influencing soil formation are temperature and moisture. Temperature has a direct influence on the weathering of bedrock to produce mineral particles. Rates of bedrock weathering generally increase with higher temperatures. Temperature also influences the activity of soil microorganisms, the frequency and magnitude of soil chemical reactions, and the rate of plant growth. Moisture levels in most soils are primarily controlled by the addition of water via precipitation minus the losses due to evapotranspiration. If additions of water from precipitation surpass losses from evapotranspiration, moisture levels in a soil tend to be high. If the water loss due to evapotranspiration exceeds inputs from precipitation, moisture levels in a soil tend to be low. High moisture availability in a soil promotes the weathering of bedrock and sediments, chemical reactions, and plant growth. The availability of moisture also has an influence on soil pH and the decomposition of organic matter.

Living Organisms have a role in a number of processes involved in pedogenesis including organic matter accumulation, profile mixing, and biogeochemical nutrient cycling. Under equilibrium conditions, vegetation and soil are closely linked with each other through nutrient cycling. The cycling of nitrogen and carbon in soils is almost completely controlled by the presence of animals and plants. Through litter fall and the process of decomposition, organisms add humus and nutrients to the soil which influences soil structure and fertility. Surface vegetation also protects the upper layers of a soil from erosion by way of binding the soils surface and reducing the speed of moving wind and water across the ground surface.

Parent Material refers to the rock and mineral materials from which the soils develop. These materials can be derived from residual sediment due to the weathering of bedrock or from sediment transported into an area by way of the erosive forces of wind, water, or ice. Pedogenesis is often faster on transported sediments because the weathering of parent material usually takes a long period of time. The influence of parent material on pedogenesis is usually related to soil texture, soil chemistry, and nutrient cycling.

Topography generally modifies the development of soil on a local or regional scale. Pedogenesis is primarily influenced by topography's effect on microclimate and drainage. Soils developing on moderate to gentle slopes are often better drained than soils found at the bottom of valleys. Good drainage enhances an number of pedogenic processes of illuviation and eluviation that are responsible for the development of soil horizons. Under conditions of poor drainage, soils tend to be immature. Steep topographic gradients inhibit the development of soils because of erosion. Erosion can retard the development through the continued removal of surface sediments. Soil microclimate is also influenced by topography. In the Northern Hemisphere, south facing slopes tend to be warmer and drier than north facing slopes. This difference results in the soils of the two areas being different in terms of depth, texture, biological activity, and soil profile development.

Time influences the temporal consequences of all of the factors described above. Many soil processes become steady state overtime when a soil reaches maturity. Pedogenic processes in young soils are usually under active modification through negative and positive feedback mechanisms in attempt to achieve equilibrium.

Principal Pedogenic Processes

A large number of processes are responsible for the formation of soils. This fact is evident by the large number of different types of soils that have been classified by soil scientists. However, at the macro-scale we can suggest that there are five main principal pedogenic processes acting on soils. These processes are laterization, podzolization, calcification, salinization, and gleization.

Laterization is a pedogenic process common to soils found in tropical and subtropical environments. High temperatures and heavy precipitation result in the rapid weathering of rocks and minerals. Movements of large amounts of water through the soil cause eluviation and leaching to occur. Almost all of the byproducts of weathering, very simple small compounds or nutrient ions, are trans located out of the soil profile by leaching if not taken up by plants for nutrition. The two exceptions to this process are iron and aluminum compounds. Iron oxides give tropical soils their unique reddish coloring. Heavy leaching also causes these soils to have an acidic pH because of the net loss of base cations.

Podzolization is associated with humid cold mid-latitude climates and coniferous vegetation. Decomposition of coniferous litter and heavy summer precipitation create a soil solution that is strongly acidic. This acidic soil solution enhances the processes of eluviation and leaching causing the removal of soluble base cations and aluminum and iron compounds from the A horizon. This process creates a sub-layer in the A horizon that is white to gray in color and composed of silica sand.

Calcification occurs when evapotranspiration exceeds precipitation causing the upward movement of dissolved alkaline salts from the groundwater. At the same time, the movement of rain water causes a downward movement of the salts. The net result is the deposition of the trans located cations in the B horizon. In some cases, these deposits can form a hard layer called caliche. The most common substance involved in this process is calcium carbonate. Calcification is common in the prairie grasslands.

Salinization is a process that functions in the similar way to calcification. It differs from calcification in that the salt deposits occur at or very near the soil surface. Salinization also takes place in much drier climates.

Gleization is a pedogenic process associated with poor drainage. This process involves the accumulations of organic matter in the upper layers of the soil. In lower horizons, mineral layers are stained blue-gray because of the chemical reduction of iron.

Soil Morphology

Soil morphology is defined as the branch of soil science that deals with the description, using standard terminology, of in situ spatial organization and physical properties of soil regardless of potential use. Precise descriptions, using conventional terms, are necessary to all of the areas of science included in the field of soil science. The basic descriptors used today have been developed over the past 50 years and are continually added to. Prior to utilization of standard terminology, soils were described as clayey, sandy, stony, sedimentary, saline, marshy, dry or moist, heavy or light, soft or compact, fatty, friable, or lean. Soil horizons were described by terms such as 'gray watery sand' or 'rusty brown clay.' While these terms illustrate a fundamental understanding of soil properties, they do not impart any specific knowledge about the soils in question and cannot be compared with other soils described by other scientists. An objective, complete description of the soil is essential, because it serves as a basis for soil identification, classification, correlation, mapping, and interpretation.

Micromorphology

While soil micromorphology begins in the field with the routine and careful use of a 10x hand lens, much more can be described by careful description of thin sections made of the soil with the aid of a petrographic polarizing light microscope. The soil can be impregnated with an epoxy resin, but more commonly with a polyester resin (crystic 17449) and sliced and ground to 0.03 millimeter thickness and examined by passing light through the thin soil plasma.

Porosity

Porosity of topsoil is a measure of the pore space in soil which typically decreases as grain size increases. This is due to soil aggregate formation in finer textured surface soils when subject to soil biological processes. Aggregation involves particulate adhesion and higher resistance to compaction. Typical bulk density of sandy soil is between 1.5 and 1.7 g/cm^3. This calculates to a porosity between 0.43 and 0.36. Typical bulk density of clay soil is between 1.1 and 1.3 g/cm^3. This calculates to a porosity between 0.58 and 0.51. This seems counterintuitive because clay soils are termed heavy, implying lower porosity. Heavy apparently refers to a gravitational moisture content effect in combination with terminology that harkens back to the relative force required to pull a tillage implement through the clayey soil at field moisture content as compared to sand.

Porosity of subsurface soil is lower than in surface soil due to compaction by gravity. Porosity of 0.20 is considered normal for unsorted gravel size material at depths below the biomantle. Porosity in finer material below the aggregating influence of pedogenesis can be expected to approximate this value.

Soil porosity is complex. Traditional models regard porosity as continuous. This fails to account for anomalous features and produces only approximate results. Furthermore it cannot help model the influence of environmental factors which affect pore geometry. A number of more complex models have been proposed, including fractals, bubble theory, cracking theory, Boolean grain process, packed sphere, and numerous other models.

Active Layer

The active layer is that portion of the soil above permafrost that thaws and freezes seasonally. It plays an important role in cold regions because most ecological, hydrological, biogeochemical and pedogenic activity takes place within it. Changes in active layer thickness are influenced by many factors, including surface temperature, physical and thermal properties of the surface cover and substrate, vegetation, soil moisture and duration and thickness of snow cover. The inter-annual and spatial variations in thaw depth at point locations can be large, an artefact of year-to-year and micro topographic variations in both surface temperature and soil moisture, and so presents monitoring challenges. When the other conditions remain constant, changes in active layer thickness could be expected to increase in response to climate warming, especially in summer.

Steps Showing the Thawing and Freezing Cycle

Under the tundra in the northern part of the Arctic Refuge, the active layer begins to thaw after the winter snows melt in early summer. Snow can fall in any month, but it doesn't usually stay on the ground until September. What date do you think the active layer is melted deepest?

Here's how the active layer changes throughout the year (These dates are averages. The actual dates on any year may be slightly earlier or later.)

On May 15, the tundra soil is still buried under winter snow, and the active layer is completely frozen.

By June 1, the snow has begun to melt, but the active layer remains frozen.

On June 15, the active layer is already thawed half way (50%) to its maximum depth. It is about 9 inches (23 cm) deep. The sun remains above the horizon for 24 hours each day, and plants are producing new growth.

By July 1, the active layer has thawed to 75% of its total depth. It is about 13.5 inches (34.5 cm) deep.

On July 15, the active layer has thawed to 85% of its total depth. It is about 15 inches (38 cm) deep.

On August 1, the active layer has thawed to 90% of its total depth. It is about 16 inches (40.5 cm) deep.

By August 15, the active layer has thawed to 92% of its total depth. It is about 16.5 inches (42 cm) deep. Autumn has come to the tundra. The air is cooler, and plants are finishing their growth for the year.

On September 1, the active layer has thawed to 94% of its total depth. It is about 16.75 inches (42.5 cm) deep. Air temperatures are now below freezing, and the soil surface begins to refreeze.

On September 15, the active layer has thawed to 96% of its total depth. It is about 17.25 inches (43.25 cm) deep. The soil surface continues to freeze down into the active layer.

By October 1, the active layer has thawed to 98% of its total depth. It is about 17.75 inches (44.5 cm) deep.

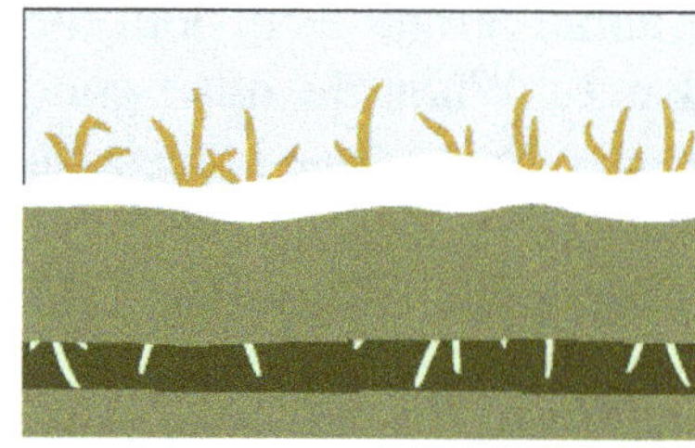

On October 15, the active layer has thawed to its maximum depth of 18 inches (about 45.5 cm). The soil above it is refreezing rapidly. Did you guess that the active layer would reach its maximum depth in mid-October?

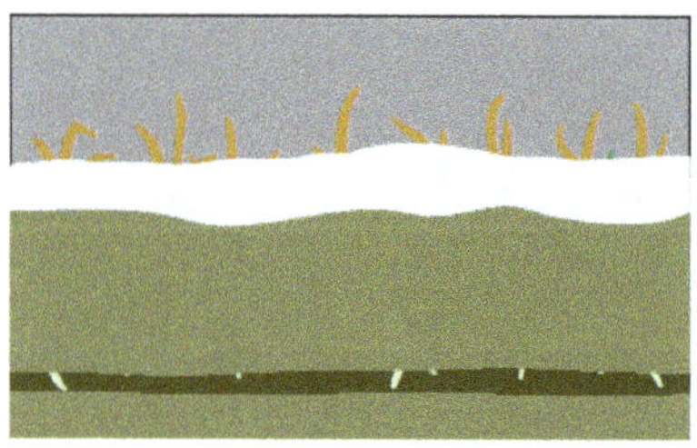

By November 1, there is just a narrow region above the base of the active layer that is not frozen.

By November 15, the active layer has completely frozen, and will remain this way until early summer.

Soil Map

Purpose of Soil Maps

A map does not represent the exact reality of the land charted, but instead provides a structured representation of knowledge about the distribution of soils across the landscape. It is an approximation of the reality of the land, and one whose accuracy increases at higher densities of observation and more detailed scales. Soil maps of a given area constitute a visualization of the intrinsic properties of the soils in question and a delimitation of their specific areas of distribution. The division of a given landscape into supposedly homogenous areas constitutes a modeling exercise which represents the distribution of soils in the landscape. Mapping this information involves a considerable effort of synthesis and concretion. A soil map should harmoniously combine the working scale, the amount of information that needs to be represented, the quality of definition and unit of delimitation and the interpretative capacity of those who use this information.

Map Provided by the DAR

The Purposes of Soil Maps are therefore to:

- Synthesize the key properties of the soils in a single document using a set of codifications and legends that allow a rapid interpretation of the results.
- Describe the distribution of soil units, grouped under more or less generalized or detailed concepts, at a suitable working scale.
- Provide a graphic document which synthesizes the inventory and/or evaluation of the soil units and their corresponding distribution over the land. con la distribución correspondiente encima del territorio.
- Contribute to the dissemination of knowledge about soils and their spatial distribution and properties. The soil map is often the only document allows a non-specialist access to soil information.
- Allow rapid access to soil information. As a specific, visual, synthetic and bi-dimensional representation of the soils of a specific area, the soil map should facilitate access to information which will enable rational use of the soil data provided.

References

- What-is-pedology: academicroom.com, Retrieved 06, March 2020
- Pedology, science: britannica.com, Retrieved 14, April 2020
- Buol, Stanley W.; Southard, Randal J.; Graham, Robert C.; McDaniel, Paul A. (2003). Soil Genesis and Classification, 5th Edition. Ames, Iowa: Iowa State Press, A Blackwell Pub. Co. p. 494. ISBN 0-8138-2873-2
- Soil-morphology, agricultural-and-biological-sciences: sciencedirect.com, Retrieved 21, July 2020

6

Edaphology and its Subfields

The branch of soil science which includes the ecological study of soil and its influence on living things, especially plants is known as edaphology. This chapter covers topics like agricultural soil science, environmental soil science, bioeffector, plant nutrition and phosphate solubilizing bacteria. The aim of this chapter is to explore the field of edaphology.

Edaphology is one of two soil sciences which also consist of Pedology. Specifically, edaphology is the study in how soils influence and interact on/with living things, especially those of plants. It is the study of the ecological relationship soil has with land cultivation practices and plants. For the cultivation of land, edaphology also focuses on the conservation of soils and possible loss or erosion due to certain practices. Viticulturists often rely on edaphologists to assist and advise on the initial planting and planning of a vineyard including vine species selection for the type of soil available.

In the study of soil science, soil is considered to have three dimensions; breadth, length and depth. Edaphology further studies the composition of soil which makes up its overall volume; mineral mater, organic matter, soil water and soil air. It also makes note that soil is comparable to the systems of animals essential to its life:

- Respiratory system - how the air circulates and exchanges gases
- Digestive system - the decomposition of organic matter
- Circulatory system - how water moves in the soil
- Excretory system - the leaching out of excess salt
- The brain - composition of clay
- Color
- Height - the soil's depth

Within viticulture, the study of soil and how it co-exists with organic life is essential to growing prosperous grape crops and staying ahead of evolutionary vine changes and climate changes.

Agricultural Soil Science

Agricultural soils are managed systems, to serve human purposes. First purpose is providing the populations with food. Second is the production of plants of high quality, as well as animal feed with sufficient protein content. Furthermore, from the economic point of view, agriculture needs yields providing sufficient income. The essential plant nutrient N has to be used appropriately for these several goals. In a longer time appreciation, soil fertility is also to be considered, soil organic matter being the main reserve of plant nutrients. Providing growing plants with enough and especially with adequate amounts of N needs (i) the knowledge about the demand of the special crop, (ii) considerations on the organic N pool in soil and (iii) on its transformation in plant-available inorganic forms (NH_4^+ and NO_3^-) as well as (iv) taking in balance the potential biological N_2 fixation and the N deposition. The mineralization of N from the organic pool is provided by soil microorganisms, using soil organic matter as source of energy and nutrient. Weather conditions strongly affect the N transformations in soil. Because weather conditions are not predictable with enough precision, the calculation of N flow from the immobile into the plant-available form is generally a site-specific determination relying on the average weather data from the last few years. As an almost general rule, NO_3^- accumulates periodically in soils. However, the main accumulation of NO_3^- occurs (i) after application of certain fertilizers or manure, (ii) after harvest, when plant residues are degraded by soil microorganisms and (iii) in late winter before plant growth. When no growing plants take up the NO_3^-, it may be denitrified or water transferred to deeper soil parts and to groundwater or surface water.

Environmental Soil Science

Environmental soil science is the study of the interaction of humans with the pedosphere as well as critical aspects of the biosphere, the lithosphere, the hydrosphere, and the atmosphere. Environmental soil science addresses both the fundamental and applied aspects of the field including: buffers and surface water quality, vadose zone functions, septic drain field site assessment and function, land treatment of wastewater, storm water, erosion control, soil contamination with metals and pesticides, remediation of contaminated soils, restoration of wetlands, soil degradation, nutrient management, movement of viruses and bacteria in soils and waters, bioremediation, application of molecular biology and genetic engineering to development of soil microbes that can degrade hazardous pollutants, land use, global warming, acid rain, and the study of anthropogenic soils, such as terra preta. Much of the research done in environmental soil science is produced through the use of models.

Bioeffector

The use of bioeffectors, formally known as plant biostimulants, has become common practice in agriculture and provides a number of benefits in stimulating growth and protecting against stress. A biostimulant is loosely defined as an organic material and/or microorganism that are applied to enhance nutrient uptake, stimulate growth, and enhance stress tolerance or crop quality.

Plant biostimulants, sometimes referred to as agricultural biostimulants, are a diverse classification of substances that can be added to the environment around a plant and have positive effects on plant growth and nutrition, but also on abiotic and biotic stress tolerance. Although most plant biostimulants are added to the rhizosphere to facilitate uptake of nutrients, many of these also have protective effects against environmental stress such as water deficit, soil salinization and exposure to sub-optimal growth temperatures. Biostimulants are not nutrients per se; instead they facilitate the uptake of nutrients or beneficially contribute to growth promotion or stress resistance. A newly emerged paradigm emphasizes that plants are not standalone entities within their environments; instead they are host and partner to microorganisms of bacteria and fungi; plants are a host to numerous micro biota and those associations, both outside and within its tissues, allow them to respond and adapt to abiotic and biotic stress. Reasonably, if we functionally optimize these associations, we may strengthen their role in plant stress protection.

The industry definition of biostimulants was originally proposed in 2012 and stated: "Plant biostimulants contain substance(s) and/or microorganisms whose function when applied to plants or the rhizosphere is to stimulate natural processes to enhance/benefit nutrient uptake, nutrient efficiency, tolerance to abiotic stress, and crop quality. Biostimulants have no direct action against pests, and therefore do not fall within the regulatory framework of pesticides". Biostimulants were loosely defined for a long time and often regarded dubiously because of their aggregate nature and the inherent difficulty to determine which specific components were making beneficial contributions. The definition proposed by du Jardin "A plant biostimulant is any substance or microorganism applied to plants with the aim to enhance nutrition efficiency, abiotic stress tolerance and/or crop quality traits, regardless of its nutrients content" represents the clearest and most concise way to define biostimulants.

Algal Extracts

Seaweed extracts (SWE) as biostimulants are emerging as commercial formulations for use as plant growth-promoting factors and a method to improve tolerance to salinity, heat, and drought. Algal extracts target a number of pathways to increase tolerance under stress. Seaweeds are red, green, and brown macro algae that represent 10% of marine productivity. Macro algae have been used as organic fertilizers for thousands of years and are still in use. Currently, there are over 47 companies producing and marketing various algal extracts for agricultural use; the majority of the formulations are from the brown algae, Ascophyllum nodosum.

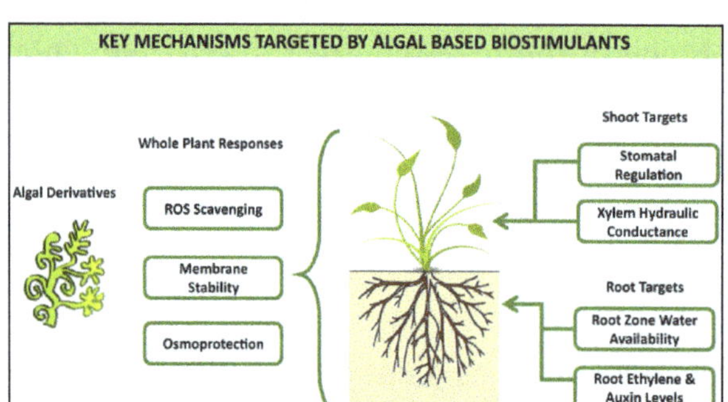

Summary of main key mechanisms targeted by algal-based biostimulants.

While the growth-promoting effects of seaweed extracts have been documented in many species, very little is actually known about the mechanisms behind these effects. The variable and complex nature of these substances makes it difficult to determine exactly which components are playing a key role. Commercial formulations of SWEs are often proprietary, and the composition is largely dependent on the method of extraction. Indeed, characterization of the actual composition of most common algal-based commercial products would be useful first step to better hypothesize and/or depict a cause–effect relationship of their mechanism of action. Mechanical disruption, pulverization, acid or alkali extractions are some of the more common methods employed. Most commercial products are derived from red (ex Lithothamnium calcareum) and brown (ex Ascophyllum nodosum, Durvillaea potatorum) macro algae. The role of SWEs and cold tolerance is now emerging. Very recent work has focused on SWEs and their ability to enhance tolerance to chilling stress. When multiple extracts were tested for their ability to enhance cold tolerance in maize only extracts rich in Zn and Mn were able to enhance tolerance through enhanced ROS responses. In this case, the protective effects likely stem from supplying plants with micronutrients that play a role as co-factors in anti-oxidative enzymes. These results indicate that nutrient deficiency stress induced by cold can be overcome by supplying SWEs rich in micronutrients to improve oxidative stress tolerance. Previous studies with corn seedlings under root chilling stress supplemented with micronutrients demonstrated the utility of nutrient seed priming.

Some work has been done in model systems with the goal of determining the physiological and molecular responses induced by SWEs. In order to better understand the active components of A. nodosum, Rayirath et al. separated the organic-sub-fractions of extracts and tested them with Arabidopsis thaliana and freezing experiments. Plants grown in vitro with sub-fractions added to the substrate or in "Peat pellet freezing assays" irrigated with sub-fractions were tested for freezing tolerance. The authors found that the ethyl acetate extracted fraction, rich in fatty acids and sterols enhanced freezing tolerance over water treated (controls) at temperatures from −2.5 to −5.5 °C. Treated plants maintained faster rates of recovery, greater membrane integrity, and had 70% less chlorophyll damage upon freezing recovery as well as increased expression of key freezing tolerance genes such as RD29A, COR15A, and CBF3. Priming of key tolerance genes prior to exposure to stress greatly increases tolerance in many cases. The lipophilic components were found to be rich in fatty acids such as butyric acid, palmitic acid, oleic acid, linoleic acid the sterol fucosterol. These extracts increased proline content and total soluble sugars, contributing to freezing tolerance. A. nodosum extracts have even been used to reduce cold stress sensitivity in Kappaphycus alvarezii. Kappaphycus alvarezii is a red algae and the most important source of carrageenan; which are hydrophilic colloids largely used in foods and dairy products. Algal extracts have also been used on Kentucky bluegrass (Poa pratensis L. cv. Plush) to alleviate salinity stress from saline watering in turf grass experiments. Similarly SWE-based cytokines have been used on creeping bent grass (Agrostis stolonifera L.) to improve tolerance to heat stress. SWEs from A. nodosum have also been used for ornamental plants, such as Spiraea nipponica"Snowmound" and Pittosporum eugenioides "Variegatum", to enhance drought tolerance. Treated plants showed higher phenolic, proline, and flavonoid content while demonstrating improved physiology under mild drought stress conditions.

In horticultural crops and trees, SWE have been largely used for similar purposes. A. nodosum SWE increased RWC, Fresh Weight, and Dry Weight in spinach (Spinacia oleracea L.) plants under drought stress with some adverse effects on the nutritional value through reduced ferrous ion chelating ability. SWE applied to seedlings of lettuce (Lactuca sativa L.) enhanced cotyledon growth similar to fertilization with potassium.

Foliar application of marine bioactive substances (isopropanol extracts from microalgae) to grape plants (Vitis vinifera L) increased leaf water potential and stomatal conductance under drought stress. Consistent with an improved stomatal response, it was also observed that K^+ and Ca^{2+} fluxes at the stomatal level were higher in treated plants. Commercial formulations of A. nodosum have been tested on almond plants (Prunus dulcis [Mill.] D. A. Webb), which demonstrated increased growth and accumulation of K^+. In conditions with ample K^+ both MegaFol and GroZyme (Valagro, Atessa, Chieti, ITALY) increased leaf area and number of leaves greater than controls treated with water or K^+. In K^+-deficiency conditions only MegaFol and a foliar application of K^+ was able to stimulate growth, although at lower levels than observed with adequate K^+ nutrition. Accumulation of K+ is an essential step in protecting against both ionic and osmotic stress and may contribute to tolerance. Orange trees, Citrus sinensis L., subjected to drought stress and treated with commercial extracts of A. nodosum had better water relations and increased water use efficiency (WUE) under irrigation at 50% restitution of evapo-transpired water. The promise of biostimulants to increase drought tolerance and WUE holds great potential for drought prone regions where horticultural crops and fruit trees are agronomically important but water availability is becoming less reliable due to urbanization and climate change.

As earlier noted, almost all of the above-mentioned experiments with SWE use commercial formulations. This may be of some concern, due to the variable nature of these products and formulation methods. A recent transcriptomic study using A. thaliana plants treated with two different commercial A. nodosum extracts showed that not all extracts are alike. One commercial product resulted in dysregulation of 4.47% of the transcriptome while the other extract only affected 0.87%. Since transcriptional priming is likely a key component in enhancing abiotic stress tolerance using SWEs, these differences imply significant variability in responses elicited. Compositions of the extracts differed greatly, indicating that choice of commercial product may have a significant effect on plant responses. Commercial formulations are often proprietary and the exact composition and extraction methods, shifting the burden to the research community to analyze and isolate the active components in these products. In order to identify and characterize how these SWEs affect plants, some form of standardization is necessary.

Carbohydrates, Proteins, Amino Acids, and Lipids

Protein hydrolysates are mixtures of polypeptides, oligopeptides, and free amino acids derived from partial hydrolysis of agricultural by-products from animals and plants. Carbohydrates, proteins, amino acids, and lipids may increase stress tolerance through different (Fig.). The effects of amino acids on ion fluxes across membranes have been clearly established, with most having a positive effect on reducing NaCl-induced potassium efflux. Protein hydrolysates (PH) are often sold as formulations that include plant growth regulators. The bulk of PH products, over 90%, are produced from chemical hydrolysis of animal by-products while enzymatically processed plant-based products are a recent development.

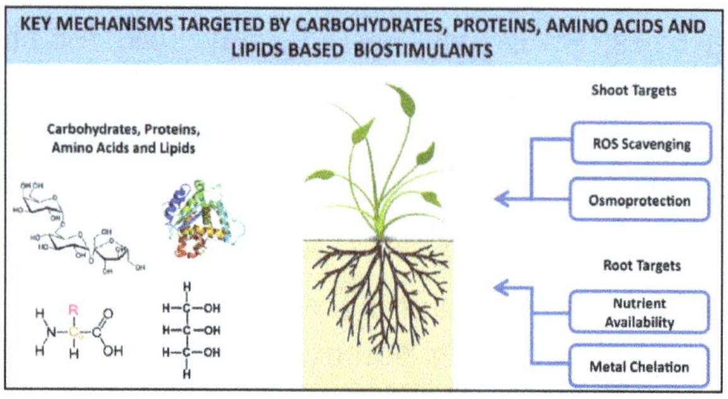

Summary of main key mechanisms targeted by carbohydrate-, protein-, amino acid-, and lipid-based biostimulants.

Megafol (Valagro, Atessa, Chieti, ITALY) is a commercial biostimulant comprising vitamins, amino acids, proteins, and betaines from plant and algal extracts. Application of Megafol to tomato plants under drought stress enhanced induction of a number of drought responsive genes such as tomato orthologs of RAB18 and RD29B. Treated plants also had higher fresh weight and relative water content under drought stress, indicating a protective effect on water status and stress responsive genes. When hydrolysate-based biostimulants from alfalfa (Medicago sativa L.), containing triacontanol (TRIA) and indole-3-acetic acid (IAA), were applied to maize plants under salt stress, the protective effects were amplified. Treated plants had higher flavonoid, proline, and potassium content in salt stress conditions over untreated controls. Extracts that are rich in amino acids may play a role in increasing cold tolerance. When lettuce plants (Lactuca sativa) were treated with an amino acid mixture, derived from enzymatic hydrolysis of proteins, (Terra-Sorb) and subjected to cold, treated plants had higher fresh weights and improved stomatal conductance. Use of animal derived amino acid hydrolysates on strawberry plants after transplantation and cold stress did not improve survival though some growth promotions were observed in the absence of stress. Perennial Rye-grass (Lolium perenne L.) treated with hydrolyzed amino acids and high temperatures (36 °C) had improved photosynthetic efficiency over control plants.

Mutants of A. thaliana deficient in production of proline have stress sensitive phenotypes. These plants can have their phenotype rescued with exogenous application of l-proline, a common amino acid available in biostimulant formulations of various amino acids and hydrolysate mixtures. Hydrolysates from wheat germs show strong anti-oxidant and free radical scavenging properties as well as the ability to chelate some metals.

Lettuce (Lactuca sativa L.) is particularly salt sensitive and the addition of plant-derived protein hydrolysates improved fresh yield, dry biomass, and root dry weight as well as increased concentrations of osmoyltes, glucosinolates and the composition of sterols and terpenes. Hydrolysates have applications for trees, which require considerable investment costs and can be vulnerable to drought. Japanese persimmon trees, Diospyros kaki L. cv. “Rojo Brillante” grafted on Diospyros lotus L., are particularly sensitive to drought stress. Treatment of these trees with calcium protein hydrolysates decreased chloride uptake under saline irrigation, lowered water potentials as well as increased the concentration of compatible solutes, all of which would enhance plant growth under saline stress.

Recent reports indicate that melatonin, derived from l-tryptophan via the shikimate pathway, can prime seeds to tolerate adverse environmental conditions at imbibition and germination stages. Corn seeds pre-treated with melatonin show increased tolerance to chilling stress upon germination, indicating a priming effect by melatonin. Melatonin may prove to be an effective biostimulant for improving stress tolerance of seedlings.

Glycinebetaine is a compatible solute accumulated in many plants in response to salt stress. Exogenous application of glycinebetaine has increased tolerance for environmental stresses such as drought, chilling, freezing, salinity, and oxidative stress. Foliar application of glycinebetaine results in rapid uptake by leaves and concentration in meristematic tissues. Rapid uptake and localization of glycinebetaine in these most vulnerable tissues are particularly beneficial in chilling and freezing stress where glycinebetaine can exert a protective effect. Transgenic plants of various species expressing two biosynthetic genes, codA and betA, produce more glycinebetaine and had an increased tolerance to abiotic stress. Exogenous application of small amounts of compatible solutes such as proline and betaine to barley roots resulted in an immediate reduction of NaCl-induced efflux of K+, indicating that ion fluxes across the membrane can be affected by relatively low concentrations of compatible solutes. The cause–effect relationship between accumulation of compatible solutes and stress protection still remains to be fully understood. However, a better understanding of the specific mechanisms of action of these molecules is becoming increasingly important if we want to make predictions on which combination of biostimulants can be more effective.

Humic and Fulvic Acids

Humic and fulvic substances are the major organic components of lignites, soil, and peat. Humic and fulvic acids are produced by the biodegradation of organic matter resulting in a mixture of acids containing phenolate and carboxyl groups. Fulvic acids are humic acids with a higher oxygen content and lower molecular weight. A number of examples exist indicating the potential for these substances to improve abiotic stress tolerance in plants (Fig.). Pre-treatment of tall fescue (Festuca arundinacea Schreb.) and creeping bent grass (Agrostis palustris Huds. A.) with seaweed extract and humic acid increased leaf hydration under dry soil conditions as well as root growth, shoot growth, and antioxidant capacity. Further studies with bent grass showed these extracts, high in cytokinins, combined with humic acid increased drought tolerance as well as endogenous cytokinin content.

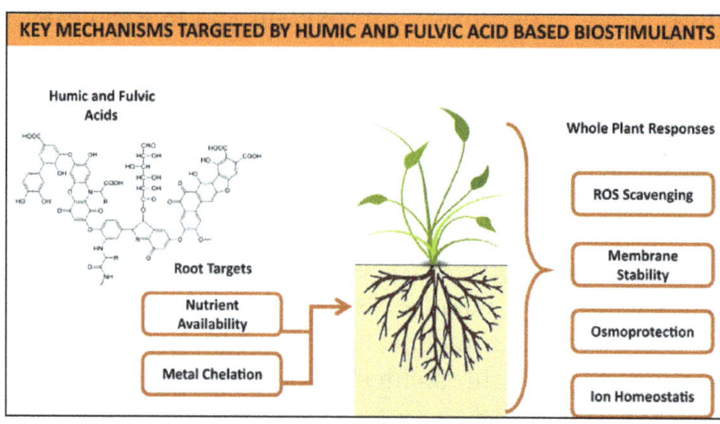

Summary of main key mechanisms targeted by humic- and fulvic acid-based biostimulants.

Treatment of bell pepper (Capsicum annuum L. cv. Demre) with humic acid and phosphorous resulted in plants with reduced Na content and elevated N, P, K, Ca, Fe, Mg, S, Mn, and Cu ion contents in roots and shoots, which were associated with a general protective effect under mild salinity stress. Application of humic acids to common bean (Phaseolus vulgaris L.) under high salinity (120 mM NaCl) increased endogenous proline levels and reduced membrane leakage, which are both indicators of better adaptation to saline environments.

Humic acid extracts seem to be beneficial also for field crop monocots. Extracts from vermicompost applied to rice (Oryza sativa L.) played a role in activating anti-oxidative enzymatic function and increased ROS scavenging enzymes. These enzymes are required to inactivate toxic-free oxygen radicals produced in plants under drought and saline stress. One possible mode of action for vermicompost may be the differential regulation of proton ATPases located in the vacuolar and plasma membranes. When Micro-Tom tomato plants were treated with vermicompost, plasma membrane proton extrusion was increased by over 40% which facilitated acid growth and nutrient uptake potential. Interestingly, the auxin insensitive mutant diageotropica (dgt) showed no increase in proton extrusion, indicating that humic substance may increase root growth through mediating auxin signalling.

Microorganisms Affecting Stress Tolerance

While plants are known to establish symbiotic relationships with bacteria, our understanding of those relationships under abiotic stress is rudimentary. However, some of the targets of microorganisms that increase abiotic stress tolerances have been identified (Fig.). Bacteria with the potential to act as biostimulants have been isolated from a number of ecosystems with saline, alkaline, acidic, and arid soils. These bacteria belong to several genera such as Rhizobium, Bradyrhizobium, Azotobacter, Azospirillum, Pseudomonas, and Bacillus. Members of these genera have developed strategies to adapt and thrive under adverse conditions. Amongst these adaptations, alterations to the composition of the cell wall and the ability to accumulate high concentrations of soluble solutes are common. These allow for enhanced water retention and increased tolerance to osmotic and ionic stress. Cell wall composition is altered through enrichment for exopolysaccharides (EPS) and lipopolysaccharide–proteins and polysaccharide–lipids which my form a protective biofilm on the root surface. Plant growth-promoting rhizobacteria (PGPR) inoculated soils can ameliorate plant abiotic stress responses. A number of recent reviews have extensively covered the protective effects of Rhizobium against abiotic stress in plants. Most documented growth enhancement determined by these bacteria is associated with high level of IAA, which has been proven to alleviate salt stress and EPS production that may help in maintaining a film of hydration around the roots and/or help re-establishing favorable water potential gradients under water limitations. These functions have been proven useful under saline stress, extremes of temperature, pH, salinity, and drought. Inoculation of maize with Azotobacter strains has been shown to have general positive effects under saline stress by facilitating uptake of K+ and exclusion of Na+ as well as increasing phosphorous and nitrogen availability. In wheat, inoculation of salt tolerance Azobacter strains increased biomass, nitrogen content, and grain yield under salt stress.

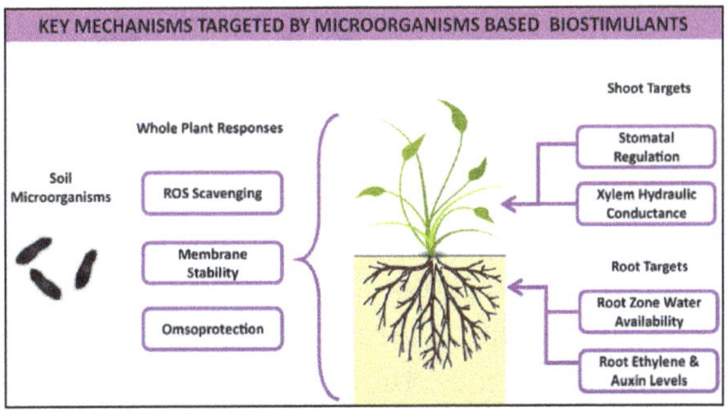

Summary of main key mechanisms targeted by microorganism-based biostimulants.

Tolerance to salt stress varies within these microorganisms and their tolerance can confer advantages to the host relationship under stress conditions. When two legumes, pea (Pisum sativum) and fava bean (Vicia faba), were inoculated with two different strains of Rhizobium leguminosarum, a salt-tolerant (GRA19) and salt-sensitive (GRL19) plants inoculated with the salt-tolerant strain performed better under moderate salt stress. The authors further found that pea plants had larger nodules and high levels of nitrogen fixation under salt stress when inoculated with GRA19, the salt-tolerant strain of R. leguminosarum. Similar results have been observed for non-symbiotic free-living soil bacteria that are capable of fixing nitrogen. Azospirillum brasilense is closely associated with the plant rhizosphere and can colonize the surface of roots. When chickpea (Cicer arietinum L.) and faba bean were inoculated with A. brasilense, they experienced enhanced nodulation by native rhizobia and greater tolerance to salt stress. Another free-living nitrogen-fixing species, Azotobacter chrococcum A2 demonstrated salt tolerance. Inoculation with A. chrococcum has been shown to increase yields of pea, potato, rice, wheat, and cotton in saline-arid soils. Increased root length and shoot growth was also observed with inoculation with significant positive yield effects for wheat (from 2.8 to 3.5 t ha^{-1} when grown in conjunction with A. chrococcum).

In barley, Hartmannibacter diazotrophicus E19 (T) is capable of colonizing roots in saline conditions. Inoculation of roots in saline soil increased root and shoot mass significantly, 308 and 189%, respectively. Inoculated roots also had increased relative water content over three and a half times that of control plants. High concentrations of salt can also be inhibitory to rhizobial bacteria. While certain strains of R. leguminosarum, such as viciae SAAN1, are very salt tolerant and able to withstand up to 0.34 M NaCl, they often show lower rates of nodulation in saline soils. These strains are often less competitive with natural rhizobial populations, however.

The stress protection of bacterial biostimulants to rain fed field crops can be of particular relevance under increasing temperatures foreseen by most climate change prediction models. Wheat inoculated with the thermo tolerant Pseudomonas putida strain AKMP7 significantly increased heat tolerance. Inoculated plants had increased biomass, shoot and root length, and seed size. ROS generation under stress treatment was also lessened, with lower levels of expression observed for ROS response genes such as superoxide dismutase, ascorbate peroxidase, and catalase. Similar results have been found with sorghum and other Pseudomonas putida AKMP strains. Psychrophilic (cold-adapted) microorganisms are capable of surviving in extreme conditions and their interactions with plants provide potential mechanisms for improving tolerance. While many strains of soil bacteria with growth-promoting properties have been isolated from low-temperature conditions, few have been tested in conjunction with plants subjected to cold stress.

Wheat inoculated with the cold-tolerant plant growth-promoting bacteria Pantoea dispersa showed improved growth and nutrient uptake, likely due to the solublization of phosphorous and production of IAA.

Inoculation of soil with psychrotolerant (cold tolerant) bacteria can play a role in chilling tolerance. The psychrotolerant soil bacterium, Burkholderia phytofirman, is a plant-growth-promoting rhizobacterium (PGPR) that is capable of colonizing multiple plant species. B. phytofirman was shown to play a role in enhancing chilling tolerance in Vitis vinifera L. by increasing ROS scavenging metabolites and stress-induced genes. Inoculated plants also recovered faster from chilling stress, returning to normal metabolic levels more quickly than controls. B. phytofirman inoculation also alters carbohydrate metabolism and accumulation while having a protective effect on net photosynthesis during cold acclimation and stress.

Tomato plants (Solanum lycopersicum cv Mill) were inoculated with cold-tolerant strains of Pseudomonas vancouverensis, and frederiksbergensis as well as Flavobacterium glaciei that were isolated from agricultural fields during winter. Treated tomato seedlings were subjected to a week of chilling stress at 15C and inoculation three of these strains showed significantly reduced electrolyte leakage and ROS activity. Improved stress tolerance and growth-promoting effects of microorganism treatments have been seen in other species also. Inoculation of lettuce (Lactuca sativa L., cv Mantecosa) seeds with A. brasilense increased germination in the presence of salt and demonstrated tolerance through higher total fresh and dry weights of plants at harvest. Additional experiments studying these effects have shown increased biomass, chlorophyll, ascorbic acid content, antioxidant content, and post-harvest shelf life after being subjected to salt stress. Sweet pepper (Capsicum annuum L.) inoculated with A. brasilense and Pantoea dispersa was not affected by moderate levels of salinization, up to 80 mM NaCl, while uninoculated control plants demonstrated lower DW starting at 40 mM NaCl.

Triticum aestivum cv. Buck Ombú inoculated with A. brasilense sp. 245 and subjected to salt stress (320 mM NaCl) and osmotic stress (20 and 30% PEG 6000) had higher FW, DW, and RWC than non-inoculated controls. Analysis of phospholipids and fatty acid composition in inoculated wheat indicated that the distribution profiles of major root phospholipids are altered in inoculated plants, possibly contributing to the increased tolerance. Wheat inoculated with Azospirillum lipoferum and irrigated with 80 mM NaCl had significantly higher leaf and root dry weight than uninoculated controls.

While the mechanisms by which A. brasilense confer tolerance to osmotic stress are not clear, some evidence indicates that inoculation induces wider xylem vessels and greater hydraulic conductance. In inoculated tomato plants subjected to water stress similar changes have been observed, such as larger xylem vessel area, higher stem-specific hydraulic conductivity, and thicker stems. Pepper plants co-inoculated with A. brasilense and Pantoea dispersa accumulated more dry matter under salt stress. Inoculated plants showed higher stomatal conductance and rates of photosynthesis under salt stress. The chlorophyll concentration and efficiency of photosystem II were not affected in inoculated plants under stress conditions.

Inhibition of root growth under salt stress conditions is well documented. One of the primary causes of this inhibition is the production and perception of ethylene in the roots. Plants and PGPR both have ACC-deaminases, which possess the ability to lower the concentration of ethylene in the roots and root zone. PGPR-derived ACC-deaminases can reduce ethylene induced inhibition by reducing root zone ethylene and contribute to maintain relatively higher root-to-shoot ration, a trait that would result beneficial under water shortage.

Biostimulant treatments of agricultural crops have the potential to improve plant resilience to environmental perturbations. In order to fine-tune application rates, biostimulant-plant specificities and techniques is identified that may yield highest impact on stress protection; high priority should be given to better understanding of the causal/functional mechanism of biostimulants.

Plant Nutrition

Plants use inorganic minerals for nutrition, whether grown in the field or in a container. Complex interactions involving weathering of rock minerals, decaying organic matter, animals, and microbes take place to form inorganic minerals in soil. Roots absorb mineral nutrients as ions in soil water. Many factors influence nutrient uptake for plants. Ions can be readily available to roots or could be "tied up" by other elements or the soil itself. Soil too high in pH (alkaline) or too low (acid) makes minerals unavailable to plants.

Fertility or Nutrition

The term "fertility" refers to the inherent capacity of a soil to supply nutrients to plants in adequate amounts and in suitable proportions. The term "nutrition" refers to the interrelated steps by which a living organism assimilates food and uses it for growth and replacement of tissue. Previously, plant growth was thought of in terms of soil fertility or how much fertilizer should be added to increase soil levels of mineral elements. Most fertilizers were formulated to account for deficiencies of mineral elements in the soil. The use of soilless mixes and increased research in nutrient cultures and hydroponics as well as advances in plant tissue analysis have led to a broader understanding of plant nutrition. Plant nutrition is a term that takes into account the interrelationships of mineral elements in the soil or soilless solution as well as their role in plant growth. This interrelationship involves a complex balance of mineral elements essential and beneficial for optimum plant growth.

Essential Versus Beneficial

The term essential mineral element (or mineral nutrient) was proposed by Arnon and Stout (1939). They concluded three criteria must be met for an element to be considered essential. These criteria are: 1. A plant must be unable to complete its life cycle in the absence of the mineral element. 2. The function of the element must not be replaceable by another mineral element. 3. The element must be directly involved in plant metabolism. These criteria are important guidelines for plant nutrition but exclude beneficial mineral elements. Beneficial elements are those that can compensate for toxic effects of other elements or may replace mineral nutrients in some other less specific function such as the maintenance of osmotic pressure. The omission of beneficial nutrients in commercial production could mean that plants are not being grown to their optimum genetic potential but are merely produced at a subsistence level. This discussion of plant nutrition includes both the essential and beneficial mineral elements.

The Mineral Elements

There are actually 20 mineral elements necessary or beneficial for plant growth. Carbon (C), hydrogen (H), and oxygen (O) are supplied by air and water. The six macronutrients, nitrogen (N), phosphorus (P), potassium (K), calcium (Ca), magnesium (Mg), and sulfur (S) are required by plants in large amounts. The rest of the elements are required in trace amounts (micronutrients). Essential trace elements include boron (B), chlorine (Cl), copper (Cu), iron (Fe), manganese (Mn), sodium (Na), zinc (Zn), molybdenum (Mo), and nickel (Ni). Beneficial mineral elements include silicon (Si) and cobalt (Co). The beneficial elements have not been deemed essential for all plants but may be essential for some. The distinction between beneficial and essential is often difficult in the case of some trace elements. Cobalt for instance is essential for nitrogen fixation in legumes. It may also inhibit ethylene formation (Samimy, 1978) and extend the life of cut roses (Venkatarayappa et al., 1980). Silicon, deposited in cell walls, has been found to improve heat and drought tolerance and increase resistance to insects and fungal infections. Silicon, acting as a beneficial element, can help compensate for toxic levels of manganese, iron, phosphorus and aluminum as well as zinc deficiency. A more holistic approach to plant nutrition would not be limited to nutrients essential to survival but would include mineral elements at levels beneficial for optimum growth. With developments in analytical chemistry and the ability to eliminate contaminants in nutrient cultures, the list of essential elements may well increase in the future.

The Mineral Elements in Plant Production

The use of soil for greenhouse production before the 1960s was common. Today a few growers still use soil in their mixes. The bulk of production is in soilless mixes. Soilless mixes must provide support, aeration, nutrient and moisture retention just as soils do, but the addition of fertilizers or nutrients are different. Many soilless mixes have calcium, magnesium, phosphorus, sulfur, nitrogen, potassium and some micronutrients incorporated as a pre-plant fertilizer. Nitrogen and potassium still must be applied to the crop during production. Difficulty in blending a homogenous mix using pre-plant fertilizers may often result in uneven crops and possible toxic or deficient levels of nutrients. Soilless mixes that require addition of micro and macronutrients applied as liquid throughout the growth of the crop, may actually give the grower more control of his crop. To achieve optimum production, the grower can adjust nutrient levels to compensate for other environmental factors during the growing season. The absorption of mineral ions is dependent on a number of factors in addition to weather conditions. These include the cation exchange capacity or CEC and the pH or relative amount of hydrogen (H+) or hydroxyl ions (OH-) of the growing medium, and the total alkalinity of the irrigation water.

CEC or Cation Exchange Capacity

The Cation Exchange Capacity refers to the ability of the growing medium to hold exchangeable mineral elements within its structure. These cations include ammonium nitrogen, potassium, calcium, magnesium, iron, manganese, zinc and copper. Peat moss and mixes containing bark, sawdust and other organic materials all have some level of cation exchange capacity.

Meaning of pH

The term pH refers to the alkalinity or acidity of a growing media water solution. This solution consists of mineral elements dissolved in ionic form in water. The reaction of this solution whether it is acid, neutral or alkaline will have a marked effect on the availability of mineral elements to plant roots. When there is a greater amount of hydrogen H+ ions the solution will be acid (<7.0). If there is more hydroxyl OH- ions the solution will be alkaline (>7.0). A balance of hydrogen to hydroxyl ions yields a pH neutral soil (=7.0). The range for most crops is 5.5 to 6.2 or slightly acidic. This creates the greatest average level for availability for all essential plant nutrients. Extreme fluctuations of higher or lower pH can cause deficiency or toxicity of nutrients.

The Elements of Complete Plant Nutrition

The following is a brief guideline of the role of essential and beneficial mineral nutrients that are crucial for growth. Eliminate any one of these elements, and plants will display abnormalities of growth, deficiency symptoms, or may not reproduce normally.

Macronutrients

Nitrogen is a major component of proteins, hormones, chlorophyll, vitamins and enzymes essential for plant life. Nitrogen metabolism is a major factor in stem and leaf growth (vegetative growth). Too much can delay flowering and fruiting. Deficiencies can reduce yields, causes yellowing of the leaves and stunt growth.

Phosphorus is necessary for seed germination, photosynthesis, protein formation and almost all aspects of growth and metabolism in plants. It is essential for flower and fruit formation. Low pH (<4) results in phosphate being chemically locked up in organic soils. Deficiency symptoms are purple stems and leaves; maturity and growth are retarded. Yields of fruit and flowers are poor. Premature drop of fruits and flowers may often occur. Phosphorus must be applied close to the plant's roots in order for the plant to utilize it. Large applications of phosphorus without adequate levels of zinc can cause a zinc deficiency.

Potassium is necessary for formation of sugars, starches, carbohydrates, protein synthesis and cell division in roots and other parts of the plant. It helps to adjust water balance, improves stem rigidity and cold hardiness, enhances flavor and color on fruit and vegetable crops, increases the oil content of fruits and is important for leafy crops. Deficiencies result in low yields, mottled, spotted or curled leaves, scorched or burned look to leaves.

Sulfur is a structural component of amino acids, proteins, vitamins and enzymes and is essential to produce chlorophyll. It imparts flavor to many vegetables. Deficiencies show as light green leaves. Sulfur is readily lost by leaching from soils and should be applied with a nutrient formula. Some water supplies may contain Sulfur.

Magnesium is a critical structural component of the chlorophyll molecule and is necessary for functioning of plant enzymes to produce carbohydrates, sugars and fats. It is used for fruit and nut formation and essential for germination of seeds. Deficient plants appear chlorotic, show yellowing between veins of older leaves; leaves may droop. Magnesium is leached by watering and must be supplied when feeding. It can be applied as a foliar spray to correct deficiencies.

Calcium activates enzymes, is a structural component of cell walls, influences water movement in cells and is necessary for cell growth and division. Some plants must have calcium to take up nitrogen and other minerals. Calcium is easily leached. Calcium, once deposited in plant tissue, is immobile (non-trans locatable) so there must be a constant supply for growth. Deficiency causes stunting of new growth in stems, flowers and roots. Symptoms range from distorted new growth to black spots on leaves and fruit. Yellow leaf margins may also appear.

Micronutrients

Iron is necessary for many enzyme functions and as a catalyst for the synthesis of chlorophyll. It is essential for the young growing parts of plants. Deficiencies are pale leaf color of young leaves followed by yellowing of leaves and large veins. Iron is lost by leaching and is held in the lower portions of the soil structure. Under conditions of high pH (alkaline) iron is rendered unavailable to plants. When soils are alkaline, iron may be abundant but unavailable. Applications of an acid nutrient formula containing iron chelates, held in soluble form, should correct the problem.

Manganese is involved in enzyme activity for photosynthesis, respiration, and nitrogen metabolism. Deficiency in young leaves may show a network of green veins on a light green background similar to an iron deficiency. In the advanced stages the light green parts become white, and leaves are shed. Brownish, black, or grayish spots may appear next to the veins. In neutral or alkaline soils plants often show deficiency symptoms. In highly acid soils, manganese may be available to the extent that it results in toxicity.

Boron is necessary for cell wall formation, membrane integrity, and calcium uptake and may aid in the translocation of sugars. Boron affects at least 16 functions in plants. These functions include flowering, pollen germination, fruiting, cell division, water relationships and the movement of hormones. Boron must be available throughout the life of the plant. It is not trans located and is easily leached from soils. Deficiencies kill terminal buds leaving a rosette effect on the plant. Leaves are thick, curled and brittle. Fruits, tubers and roots are discolored, cracked and flecked with brown spots.

Zinc is a component of enzymes or a functional cofactor of a large number of enzymes including auxins (plant growth hormones). It is essential to carbohydrate metabolism; protein synthesis and internodal elongation (stem growth). Deficient plants have mottled leaves with irregular chlorotic areas. Zinc deficiency leads to iron deficiency causing similar symptoms. Deficiency occurs on eroded soils and is least available at a pH range of 5.5 - 7.0. Lowering the pH can render zinc more available to the point of toxicity.

Copper is concentrated in roots of plants and plays a part in nitrogen metabolism. It is a component of several enzymes and may be part of the enzyme systems that use carbohydrates and proteins. Deficiencies cause die back of the shoot tips, and terminal leaves develop brown spots. Copper is bound tightly in organic matter and may be deficient in highly organic soils. It is not readily lost from soil but may often be unavailable. Too much copper can cause toxicity.

Molybdenum is a structural component of the enzyme that reduces nitrates to ammonia. Without it, the synthesis of proteins is blocked and plant growth ceases. Root nodule (nitrogen fixing) bacteria also require it. Seeds may not form completely, and nitrogen deficiency may occur if plants are lacking molybdenum. Deficiency signs are pale green leaves with rolled or cupped margins.

Chlorine is involved in osmosis (movement of water or solutes in cells), the ionic balance necessary for plants to take up mineral elements and in photosynthesis. Deficiency symptoms include wilting, stubby roots, chlorosis (yellowing) and bronzing. Odors in some plants may be decreased. Chloride, the ionic form of chlorine used by plants, is usually found in soluble forms and is lost by leaching. Some plants may show signs of toxicity if levels are too high.

Nickel has just recently won the status as an essential trace element for plants according to the Agricultural Research Service Plant, Soil and Nutrition Laboratory in Ithaca, NY. It is required for the enzyme urease to break down urea to liberate the nitrogen into a usable form for plants. Nickel is required for iron absorption. Seeds need nickel in order to germinate. Plants grown without additional nickel will gradually reach a deficient level at about the time they mature and begin reproductive growth. If nickel is deficient plants may fail to produce viable seeds.

Sodium is involved in osmotic (water movement) and ionic balance in plants.

Cobalt is required for nitrogen fixation in legumes and in root nodules of non-legumes. The demand for cobalt is much higher for nitrogen fixation than for ammonium nutrition. Deficient levels could result in nitrogen deficiency symptoms.

Silicon is found as a component of cell walls. Plants with supplies of soluble silicon produce stronger, tougher cell walls making them a mechanical barrier to piercing and sucking insects. This significantly enhances plant heat and drought tolerance. Foliar sprays of silicon have also shown benefits reducing populations of aphids on field crops. Tests have also found that silicon can be deposited by the plants at the site of infection by fungus to combat the penetration of the cell walls by the attacking fungus. Improved leaf erectness, stem strength and prevention or depression of iron and manganese toxicity have all been noted as effects from silicon. Silicon has not been determined essential for all plants but may be beneficial for many.

Phosphate Solubilizing Bacteria

To meet the demand of overgrowing population it is the need of agrarian community to enhance the yield and future food supply. To overcome these problems, efforts needed to focus the soil biological system and the agro-ecosystem for better understanding the complex processes and their interactions for governing the stability of agricultural land.

The green revolution has been proved the most intellectual human activities contributing global food security and, consequently, changes the face of developing countries, such as India, from being food-deficient to having a food surplus. In present conditions there is an urgent need of second green revolution to increase the food production by around 50% in coming next 20 years to fulfill the demand of increasing population pressure.

Chemical fertilizers, such as water-soluble phosphatic (WSP) fertilizers have played a significant role in the green revolution to rectify the phosphorus deficiencies. However, excessive use of chemical pesticides arises soil health issues and beyond certain limit the yield plateau get declined. Thus it becomes clear that conventional agricultural practices cannot sustain the production base, for too long; while, to augment crop productivity agronomists have to rely on chemical fertilizers. In this context, after nitrogen, phosphorus is an essential plant nutrient whose deficiency marked the high yield. The phosphorous is present only in micro molar or lesser quantities in the earth's crust and is highly reactive with other elements in the soil.

Phosphorous is a one of the most abundant metallic elements found in the earth's crust and is present in the soils in both inorganic and organic forms. It is utilized or absorbed by the plants in inorganic form i.e. in orthophosphate (H2PO4- and HPO42-). It has a key role in metabolic processes such as photosynthesis, energy transfer, signal transduction, nitrogen fixation in legumes, crop quality and resistance to plant diseases are the main features associated with phosphorous nutrition.

Phosphorus being a structural component of many coenzymes, phospho-proteins, phospholipids also forms a part of the genetic memory "DNA" of all living organisms. It involved in transfer and storage of energy which used for growth and reproduction. Phosphorus plays a lead role in especially in photosynthesis, carbon metabolism, and membrane formation also the vital role in elongation of root, proliferation, and phosphorous deficiency affects root architecture. A major portion of phosphorus absorbed by the plant is accumulated in grain in the form of phytic acid which becomes unavailable to plants and its deficiency negatively affects grains yield.

Tropical and subtropical regime has acidic soil considered as extremely deficient in phosphorus with high phosphorus sorption (fixation) capacities. On average, most mineral nutrients in soil solution are present in millimolar amounts but phosphorus is present only in micromolar or lesser quantities. The low levels of phosphorus are due to high reactivity of soluble phosphate with other elements.

A number of heterotrophic microorganisms excreting organic acids which solubilized P that chelate cationic partners of P ions and release the P directly into solution. These phosphate solubilizing bacteria (PSB) are being used as bio fertilizer since 1950s. Microbial inoculant assimilates soluble P, and prevents it from adsorption or fixation. These microorganisms influences soil fertility through various processes viz. decomposition, mineralization and release of nutrients. Microorganisms enhance the P availability to plants through solubilization of inorganic form of P to in available form. Hence, microbial inoculants are used as an alternate source, which are both economic as well as ecofriendly. A continued exploration of the natural biodiversity of soil microorganisms and the optimization of microbial interactions in the rhizosphere represents a prerequisite step to develop the more efficient microbial inoculants with phosphorus-solubilizing ability.

Phosphate –Plant Interaction

Phosphorus is one the major nutrient limiting plant growth. It has diverse role in plant nutrition and promotes the development of deeper roots. The soil that is rich in phosphorus constitutes about 0.05% (w/w) phosphorus but only one tenth of this is available to plants. Most of the P (95-99%) present in the soil in the insoluble form and hence cannot be utilized by the plants due to chemical fixation in the soil and it's interaction with other metallic elements that are present in the rhizospheric area. To increase the availability of phosphorus for plants, large amounts of fertilizer is used on a regular basis. But the continuous application of fertilizer P is rapidly transferred to the insoluble forms and thus there is a need of phosphate solubilizing microorganism to make the P in available form to the plants.

Phosphate Solubilizing Micro-Organisms

Naturally occurring rhizospheric phosphorus solubilizing microorganism (PSM) dates back to 1903. Number of microbial species plays key role in P solubilization these includes bacteria, fungi, actinomycetes and even algae. Bacteria are predominant amongst them and proved more effective in phosphorus solubilization than fungi. In addition to Pseudomonas and Bacillus, other bacteria reported as P solubilizes these are Rhodococcus, Arthrobacter, Serratia, Chryseobacterium, Phyllobacterium etc., Azotobacter, Xanthomonas, Enterobacter, Pantoea, and Klebsiella. Several halophilic bacteria Kushneria sinocarni have also been isolated from the sediment of Daqiao saltern on the eastern coast of China, which may be useful in stress conditions; salt affected agricultural soils. Among the whole microbial population in soil, PSB constitute 1 to 50 %, while phosphorus solubilizing fungi (PSF) are only 0.1 to 0.5 % in P solubilization potential which includes Penicillium and Aspergillus, Rhizoctonia solani, Trichoderma.

Moreover, fungi in soils are able to traverse long distances more easily than bacteria and hence, may be more important to P solubilization in soils. Generally, the P-solubilizing fungi produce more acids compared to bacteria and thus lead more P-solubilizing activity. Among the yeasts, Yarrowia lipolytica has potential to solubilize phosphate. Algae such as cyanobacteria and mycorrhiza have been also reported for P solubilization activity.

Mechanism of Phosphate Solubilization

The mechanism of P solubilization that is employed mostly by soil microorganisms includes: (1) release of complex compounds e.g. organic acid anions, siderophores, protons, hydroxyl ions, CO_2, (2) liberation of extracellular enzymes or it also referred as biochemical P mineralization and (3) the release of P during the degradation of substrate. Thus, microorganisms have key role in the soil P cycle i.e. precipitation, sorption–desorption, and mineralization.

Inorganic Phosphate Solubilization

Microorganism plays an important role in P solubilization through secretion of organic acid production either by: (i) lowering the pH, or (ii) through chelation reaction of cations bound to P (iii) by competing with P for adsorption sites on the soil.

The lowering in pH of the medium suggests the secretion of organic acids by the P solubilizing microorganisms via direct oxidation pathway that occurs on the outer face of the cytoplasmic membrane. When P is applied to soil it get interact with other metallic elements such as Fe, Al and Ca ions which makes the P unavailable to plants through the formation of ferrous phosphate, aluminium phosphate, calcium phosphate etc. and the release of organic acids by PSM leads the chelation reaction and because of this the bound P to other metallic elements get freed and becomes available to plants. The prominent acids that are released by PSM in the solubilization of insoluble P are gluconic acid, oxalic acid, citric acid, lactic acid, tartaric acid and aspartic acid etc.

The another mechanism is the production of H_2S, which react with ferric phosphate to yield ferrous sulphate with the release of phosphate. It could be because of the activity of PSM occurs as a consequence of microbial sulphur oxidation, nitrate production and CO_2 formation. These processes ultimately leads the formation of inorganic acids like sulphuric acid.

Organic Phosphate Solubilization

Phosphorus can be released in the soil from organic compounds by three groups of enzymes: (1) Nonspecific phosphatases, which leads dephosphorylation of phosphoester or phosphoanhydride bonds in organic matter, (2) Phytases, which mostly release P which is intact in the form of phytic acid, and (3) Phosphonatases and C–P Lyases, the phosphonates degrading enzyme enzymes that perform C–P cleavage in organophosphates. Availability of organic phosphate compounds for plant nutrition could be a limitation because as phosphorous is highly reactive it will interact with other metallic elements that are present in the soil in the rhizospheric area and becomes unavailable to plants which retard the plant growth and subsequently crop yield. Therefore, the capability of enzymes to perform the desired function in the rhizosphere is a crucial aspect for their effectiveness in plant nutrition.

Role of Microbial Exopolysaccharides in Phosphate Solubilization

Recently the role of polysaccharides in the microbial mediated solubilization of P was assessed by Yi et al. (2008). Microbial exopolysaccharides (EPSs) are polymers that mainly consist of carbohydrates excreted by some bacteria and fungi onto the outside of their cell walls. Bacterial strains such as Enterobacter sp. (EnHy-401), Arthrobacter sp. (ArHy-505), Azotobacter sp. (AzHy-510) and Enterobacter sp. (EnHy-402), has the ability to solubilize TCP (tri calcium phosphate).

Phosphate Solubilizing Bacteria as Plant Growth Promoters

There are several reports on plant growth promotion by bacteria that have the ability of solubilize inorganic and/or organic P from soil after their inoculation in soil or plant seeds. It was reported that a strain of Burkholderia cepacia, commercially used as bio fertilizer in Cuba which display significant mineral phosphate solubilization and moderate phosphatase activity, also improve the yield of tomato, potato, onion, banana, coffee etc. Inoculation with two strains of Rhizobium leguminosarum selected for their P solubilization ability has been shown to improve root colonization and growth promotion and to increase significantly the P concentration in lettuce and maize. Also a strain of Pseudomonas putida stimulates the growth of roots and shoots and increased 32P-labeled phosphate uptake in canola. Co-inoculation of Pseudomonas striata and Bacillus polymyxa strains showing phosphate solubilizing activity, with a strain of Azospirillium brasilense, resulted in significant increase in grain and dry matter yields, with a concomitant increase in N and P uptake. Several studies have shown that PSB interacts with the vesicular arbuscular mycorrhizae (VAM) by releasing phosphate ions in the soil, which causes synergistic interaction that allows for better exploitation of poorly soluble P sources. Phylazonit-M is the commercial bio fertilizer which contains the mixtures of different bacterial cultures such as Bacillus megaterium, Azotobacter chroococcum increases N and P supply to the plants similarly, other product named "KYUSEI EM", a mixed inoculums of lactic acid bacteria, the organic acid lactic acid being the agent for the mineral phosphate solubilization.

Genetic Engineering of Phosphate Solubilizing Microorganisms

Some of the genes have been reported involved in mineral and organic phosphate solubilization has been so far isolated and characterized. Genetic manipulation of these genes followed by their expression in selected rhizobacterial strains leads a promising perspective for obtaining PSM strains with higher phosphate solubilizing capacity, and thus it become more effective as agricultural inoculants. The initial achievement was achieved by Goldstein and Liu from the Gram negative bacteria Erwinia herbicola through cloning of P solubilization genes. Similarly, the nap A phosphatase gene from the soil bacterium Morganella morganii was transferred to Burkholderia cepacia IS-16, a strain used as a bio fertilizer, using the broad host range vector pRK293. Introduction of P solubilization genes in natural rhizosphere bacteria is a candidate approach for the improvement of microbial capacity.

Fourteen different nonspecific acid phosphatase encoding genes have been isolated from different bacterial species using different expression cloning system. Sequence analysis of the cloned phosphatase genes allowed the classification into three different families: class A, class B, and class C phos- phatases. Several other phosphatase genes have been isolated from Escherichia coli. These include: ushA, which encodes a 5'-nucleotidase; agp, which encodes an acid glucose-1- phosphatase and cpdB, encoding the 2'-3' cyclic phosphodiesterase. Despite the difficulties, significant progress has been made for obtaining genetically engineered microorganisms for the agricultural use.

In conclusion, phosphorus is an essential element in crop nutrition. Continuous application of chemical based P fertilizers compelled us to find a sustainable approach for efficient P availability in agriculture to meet the over growing demand of food. Soil microorganisms plays varied role that affect the transformation of P and thus influence the availability of P to plant roots.

Table: Biodiversity of P Solubilizing Microorganisms.

Groups	Examples
N_2 fixing biofertilizers	
Free living	Azotobacter, Beijerinkia, Clostridium, Klebsiella, Anabaena and Nostoc
Symbiotic	Rhisobium, Frankia and Anabaena azollae
Associative symbiotic	Azospirillum
P solubilizing bifrertilizers	
Bacteria	Bacillus megaterium var. phosphaticum, Bacillus circulans and Pseudomonas striata
Fungi	Penicillium sp. and Aspergillus awamori
P mobilizing biofertilizers	
Arboscular mycorhiza	Glomus sp., Gigaspora sp., Acaulospora sp., and Sclerocystis sp.
Ectomycorrhiza	Laccaria sp., Boletus sp., and Amanita sp.
Ericoidmycorrhiza	Pezizella
Orchid mycorrhiza	Rhizoctonia solani
Biofertilizers for micro nutrients	
Silicate and Zinc solubilizers	Bacillus sp.
Plant growth promoting rhizobacteria	
Pseudomonas	Pseudomonas fluorescens

The use of efficient PSM (phosphatesolubilizing microorganisms), opens up a new horizon for better crop productivity and for greater yield performance without affecting the soil health. Phosphate solubilizing bacteria play an important role in the plant nutrition through increase in P uptake by the plants and their use as PGPR is an important contribution to bio fertilization of agricultural crops. Therefore, steps should be taken for extensive and consistent research for the identification and characterization of PSM with greater efficiency for their ultimate application under field conditions. It becomes the responsibility of soil scientists and microbiologists how soil P could be improved without applying the chemical based phosphatic fertilizers under different agroclimatic regions.

Genetic engineering of the phosphate solubilizing character must eventually be directed to the chromosomal integration of the gene for higher stability of the character and to avoid horizontal transfer of the inserted gene in soil. This strategy would also prevent the risk of metabolic load caused by the presence of the plasmid in the bacterial cell.

Thus the exploitation of the candidate soil microorganisms which play an important role in the mobilization of soil P and understanding the mechanism of phosphate solubilization, phosphate plant interaction and their contribution to the cycling of P in soil plant systems is essential for the development of sustainable agriculture to forward and accomplished our movement from a green revolution to an evergreen revolution.

References

- Ziadat, Feras Mousa; Yeganantham, Dhanesh; Shoemate, David; Srinivasan, Raghavan; Narasimhan, Balaji; Tech, Jaclyn (2015). “Soil-Landscape Estimation and Evaluation Program (SLEEP) to predict spatial distribution of soil attributes for environmental modeling”. International Journal of Architectural and Biological Engineering. 8 (3): 158–172. doi:10.3965/j.ijabe.20150803.1270
- Edaphology-1800: winefrog.com, Retrieved 28, January 2020
- Agricultural-soil, agricultural-and-biological-sciences: sciencedirect.com, Retrieved 19, August 2020
- Chen, Hongwei; An, Jing; Wei, Shuhe; Gu, Jian; Liang, Wenju (2015). “Spatial Patterns and Risk Assessment of Heavy Metals in Soils in a Resource-Exhausted City, Northeast China”. PLOS ONE. 10 (9): e0137694. doi:10.1371/journal.pone.0137694

7

Soil Survey

The systematic study of description, classification and mapping of soil is known as soil survey. It applies the theories of soil formation, physical geography, land use patterns and analysis of vegetation. This chapter closely examines the key concepts of soil survey including catena, liming and pedotransfer function, etc. to provide an extensive understanding of the subject.

Soil survey, or soil mapping, is the process of classifying soil types and other soil properties in a given area and geo-encoding such information. It applies the principles of soil science, and draws heavily from geomorphology, theories of soil formation, physical geography, and analysis of vegetation and land use patterns. Primary data for the soil survey are acquired by field sampling and by remote sensing. Remote sensing principally uses aerial photography, but LiDAR and other digital techniques are steadily gaining in popularity. In the past, a soil scientist would take hard-copies of aerial photography, topo-sheets, and mapping keys into the field with them. Today, a growing number of soil scientists bring a ruggedized tablet computer and GPS into the field with them. The tablet may be loaded with digital aerial photos, LiDAR, topography, soil geodatabases, mapping keys, and more.

The term *soil survey* may also be used as a noun to describe the published results. In the United States, these surveys were once published in book form for individual counties by the National Cooperative Soil Survey. Today, soil surveys are no longer published in book form; they are published to the web and accessed on NRCS Web Soil Survey where a person can create a custom soil survey. This allows for rapid flow of the latest soil information to the user. In the past it could take years to publish a paper soil survey. Today it takes only moments for changes to go live to the public. Also, the most current soil survey data is made available at NRCS Soil Data Mart for high end GIS users such as professional consulting companies and universities.

The information in a soil survey can be used by farmers and ranchers to help determine whether a particular soil type is suited for crops or livestock and what type of soil management might be required. An architect or engineer might use the engineering properties of a soil to determine whether it is suitable for a certain type of construction. A homeowner may even use the information for maintaining or constructing their garden, yard, or home.

Soil Survey Components

Typical information in a published county soil survey includes the following:

- a brief overview on how to use the survey
- a general soil map for comparing the sustainability of large sections of the county
- a detailed map with specific soil series outlined and indexed
- a section on the use and management of soils
- tables describing the physical features and environment of the county
- tables containing land use suitability based on standards set by the Natural Resources Conservation Service.

Catena

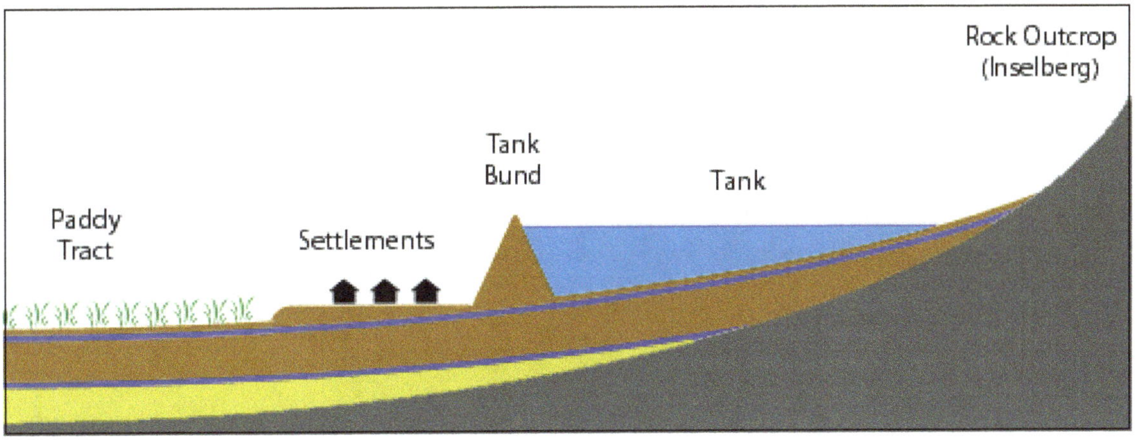

Cross Section of Dry Zone Catena of Sri Lanka showing relationship to rural land use.

A Catena in soil science (pedology) is a series of distinct but co-evolving soils arrayed a slope. Each soil type or "facet" differs somewhat from its neighbours, but all occur in the same climate and on the same underlying parent material. A mature catena is in equilibrium as the processes of deposition and erosion are in balance.

The Catena Concept

The catena concept originated in Central Uganda, by chemist W.S. Martin to describe a hill slope sequence at the Bukalasa research station. The term catena (Latin: chain) was first coined by scientist Geoffrey Milne to describe these soil-topography units. The concept of a catena was developed in order to analyze the regular variation of soils across a slope. The example of this approach consists first in a structural component, the recurring pattern of certain soils in a landscape transects in which every chain element has its place in the chain, a soil has it in a landscape.

Formation

The term soil catena is used to describe the lateral variation in soils over a hillslope. A slope can be broken into sections known as a ridge, crest, midslope, and toeslope. The ridge or hilltop tends to accumulate organic matter that allows formation of an adequate thickness of soil. Steeper slope or crest sections tend to be freely drained, while at the bottom of slopes or toeslopes there is usually higher in moisture content due to poor drainage. Toeslope soils are not only higher in moisture content, but are also known to be richer in clay and organic matter.

Lithology and relief can be the primary controls on the development of certain catenas with easily disaggregated parent rock and high relief favoring particle redistribution and therefore the formation of distinct soils in particle-source and particle-deposition zones along a slope.

Catenas can also develop on low relief hillslopes, but because less potential energy is available the redistribution of mass can be dominated by subsurface flow of plasma, a combination of dissolved and suspended solids in soil water.

Open System

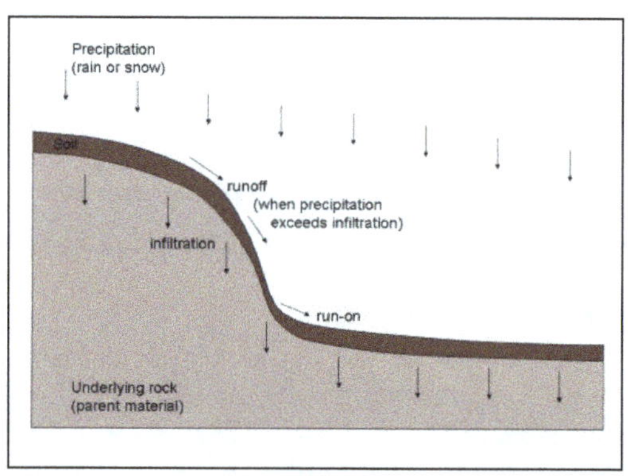

A catena is a sequence of soils down a slope, created by the balance of processes such as precipitation, infiltration and runoff.

Gullies in wet peaty soil in Scotland show where water has run off, before sinking into deeper soils at the bottom of the catena.

A catena forms when the climate, including precipitation and evaporation, is the same for the whole slope, and when sufficient time has passed for equilibrium to be reached between the processes that bring materials in to a facet and the processes that take materials away. The result is a predictable sequence of soil facets. A catena is thus an open system which has continuous input and output processes. On a steeper slope in the middle of a catena, erosion, such as by runoff, is faster, so facets are typically thinner and drier. Conversely, on a shallower slope at the top or bottom of a catena, soils are thicker and deeper. In addition, the top facets lose materials such as mineral salts when these are washed out by rain (eluviation), while the bottom facets gain materials when these are washed in (illuviation).

A catena can form on various underlying or parent materials and in different climates. On impermeable acid rocks such as metamorphic schists in a high rainfall climate like that of Western Scotland, the catena consists of thick acidic peat forming wet bog on the flatter facets, and thinner, drier, somewhat less acidic peaty podsols on the steeper facets. Thus the soil depth, acidity (pH) and soil moisture vary continuously along the slope. On a permeable basic rock such as chalk, the catena may consist of thick brown earths on the flatter facets, with thin rendzinas on the steeper slopes, while the valley bottom may include alkaline fen peat or river alluvium.

Importance

The importance of a catena is the variation of soils across a small area such as a slope. Understanding the soils that make up a catena could facilitate the mapping of soils across a given region.

Many fields of study are taken into consideration when studying catenas, studying catenas could help to understand the influence of soil hydrology on soil formation.

Catenas are found to be a great location for the study of soil science, given that the catena concept focuses on past history of the land surface, on hydrology, erosion, sediment transport, and pedogenic processes.

Liming

Liming is the application of calcium- and magnesium-rich materials to soil in various forms, including marl, chalk, limestone, or hydrated lime. This neutralises soil acidity and increases activity of soil bacteria. However, oversupply may result in harm to plant life. Lime is a basic chemical, the effect of it makes the soil more basic thus making acidic soils neutral.

Liming of a field in Devon.

The degree to which a given amount of lime per unit of soil volume will increase soil pH depends on the cation exchange capacity (CEC). Soils with low CEC will show a more marked pH increase than soils with high CEC. But the low-CEC soils will witness more rapid leaching of the added bases, and so will see a quicker return to original acidity unless additional liming is done.

Over-liming is most likely to occur on soil which has low CEC, such as sand which is deficient in buffering agents such as organic matter and clay.

Most acid soils are saturated with aluminium rather than hydrogen ions. The acidity of the soil is therefore a result of hydrolysis of aluminium. This concept of "corrected lime potential" to define the degree of base saturation in soils became the basis for procedures now used in soil testing laboratories to determine the "lime requirement" of soils.

An agricultural study at the Faculty of Forestry in Freising, Germany that compared tree stocks 2 and 20 years after liming found that liming promotes nitrate leaching and decreases the phosphorus content of some leaves.

Pedotransfer Function

In soil science, pedotransfer functions (PTF) are predictive functions of certain soil properties using data from soil surveys.

The term *pedotransfer function* was coined by Johan Bouma as *translating data we have into what we need.* The most readily available data come from soil survey, such as field morphology, soil texture, structure and pH. Pedotransfer functions add value to this basic information by translating them into estimates of other more laborious and expensively determined soil properties. These functions fill the gap between the available soil data and the properties which are more useful or required for a particular model or quality assessment. Pedotransfer functions utilize various regression analysis and data mining techniques to extract rules associating basic soil properties with more difficult to measure properties.

Although not formally recognized and named until 1989, the concept of the pedotransfer function has long been applied to estimate soil properties that are difficult to determine. Many soil science agencies have their own (unofficial) *rule of thumb* for estimating difficult-to-measure soil properties. Probably because of the particular difficulty, cost of measurement, and availability of large databases, the most comprehensive research in developing PTFs has been for the estimation of water retention curve and hydraulic conductivity.

History

The first PTF came from the study of Lyman Briggs and McLane (1907). They determined the wilting coefficient, which is defined as percentage water content of a soil when the plants growing in that soil are first reduced to a wilted condition from which they cannot recover in an approximately saturated atmosphere without the addition of water to the soil, as a function of particle-size:

$$\textit{Wilting coefficient} = 0.01\ \textit{sand} + 0.12\ \textit{silt} + 0.57\ \textit{clay}$$

With the introduction of the field capacity (FC) and permanent wilting point (PWP) concepts by Frank Veihmeyer and Arthur Hendricksen (1927), research during the period 1950-1980 attempted to correlate particle-size distribution, bulk density and organic matter content with water content at field capacity (FC), permanent wilting point (PWP), and available water capacity (AWC).

In the 1960s various papers dealt with the estimation of FC, PWP, and AWC, notably in a series of papers by Salter and Williams (1965 etc.). They explored relationships between texture classes and available water capacity, which are now known as class PTFs. They also developed functions relating the particle-size distribution to AWC, now known as continuous PTFs. They asserted that their functions could predict AWC to a mean accuracy of 16%.

In the 1970s more comprehensive research using large databases was developed. A particularly good example is the study by Hall et al. (1977) from soil in England and Wales; they established field capacity, permanent wilting point, available water content, and air capacity as a function of textural class, and as well as deriving continuous functions estimating these soil-water properties. In the USA, Gupta and Larson (1979) developed 12 functions relating particle-size distribution and organic matter content to water content at potentials ranging from -4 kPa to -1500 kPa.

With the flourishing development of models describing soil hydraulic properties and computer modelling of soil-water and solute transport, the need for hydraulic properties as inputs to these models became more evident. Clapp and Hornberger (1978) derived average values for the parameters of a power-function water retention curve, sorptivity and saturated hydraulic conductivity for different texture classes. In probably the first research of its kind, Bloemen (1977) derived empirical equations relating parameters of the Brooks and Corey hydraulic model to particle-size distribution.

Jurgen Lamp and Kneib (1981) from Germany introduced the term pedofunction, while Bouma and van Lanen (1986) used the term transfer function. To avoid confusion with the term *transfer function* used in soil physics and in many other disciplines, Johan Bouma (1989) later called it *pedotransfer function*. (A personal anecdote hinted that Arnold Bregt from Wageningen University suggested this term).

Since then, the development of hydraulic PTFs has become a boom research topic, first in the US and Europe, South America, Australia and all over the world.

Although most PTFs have been developed to predict soil hydraulic properties, they are not restricted to hydraulic properties. PTFs for estimating soil physical, mechanical, chemical and biological properties have also been developed.

Software

There are several available programs that aid determining hydraulic properties of soils using pedotransfer functions, among them are:

- SOILPAR – By Acutis and Donatelli
- ROSETTA – By Schaap et al. of the USDA, uses artificial neural networks

Soil Inference Systems

McBratney et al. (2002) introduced the concept of a soil inference system, SINFERS, where pedotransfer functions are the knowledge rules for soil inference engines. A soil inference system takes measurements with a given level of certainty (source) and by means of logically linked pedotransfer functions (organiser) infers data that is not known with minimal inaccuracy (predictor).

References

- Lindbo, David L.; Kozlowski, Deb A.; Robinson, Clay, eds. (2012). Know Soil, Know Life. Madison, Wis.: Soil Science Society of America. pp. 94–5. ISBN 9780891189541
- "Soil Survey". NRCS Nebraska. US Department of Agriculture. Retrieved 13, July 2020
- Sommer, M., Halm, D., Weller, U., Zarei, M., Stahr, K., 2000. "Lateral podzolization in a granite landscape". Soil Science Society of America Journal 64 (6), 2000. doi:10.2136/sssaj2000.6441434x
- "Corrected lime potential (formula)". Sis.agr.gc.ca. 2008-11-27. Retrieved 13, May 2020
- Acutis, M., and Donatelli, M. (2003). "SOILPAR 2.00: software to estimate soil hydrological parameters and functions". European Journal of Agronomy. 18 (3–4): 373–377. doi:10.1016/S1161-0301(02)00128-4

8

Soil Erosion and Contamination

The presence of contaminants in soil beyond a certain concentration which cause loss of functions of soil is known as soil contamination. The transportation of upper layer of soil from one place on Earth's surface to another place due to the action of water, wind, animals or human activities is known as soil erosion. This chapter has been carefully written to provide an easy understanding of the varied facets of soil loss and soil contamination.

Erosion

An actively eroding rill on an intensively-farmed field in eastern Germany.

In earth science, erosion is the action of surface processes (such as water flow or wind) that remove soil, rock, or dissolved material from one location on the Earth's crust, then transport it away to another location. The particulate breakdown of rock or soil into clastic sediment is referred to as *physical* or *mechanical* erosion; this contrasts with *chemical* erosion, where soil or rock material is removed from an area by its dissolving into a solvent (typically water), followed by the flow away of that solution.

Eroded sediment or solutes may be transported just a few millimetres, or for thousands of kilometres.

Natural rates of erosion are controlled by the action of geomorphic drivers, such as rainfall; bedrock wear in rivers; coastal erosion by the sea and waves; glacial plucking, abrasion, and scour; areal flooding; wind abrasion; groundwater processes; and mass movement processes in steep landscapes like landslides and debris flows. The rates at which such processes act control how fast a surface is eroded. Typically, physical erosion proceeds fastest on steeply sloping surfaces, and rates may also be sensitive to some climatically-controlled properties including amounts of water supplied (e.g., by rain), storminess, wind speed, wave fetch, or atmospheric temperature (especially for some ice-related processes). Feedbacks are also possible between rates of erosion and the amount of eroded material that is already carried by, for example, a river or glacier. Processes of erosion that produce sediment or solutes from a place contrast with those of deposition, which control the arrival and emplacement of material at a new location.

While erosion is a natural process, human activities have increased by 10-40 times the rate at which erosion is occurring globally. Excessive (or accelerated) erosion causes both "on-site" and "off-site" problems. On-site impacts include decreases in agricultural productivity and (on natural landscapes) ecological collapse, both because of loss of the nutrient-rich upper soil layers. In some cases, the eventual end result is desertification. Off-site effects include sedimentation of waterways and eutrophication of water bodies, as well as sediment-related damage to roads and houses. Water and wind erosion are the two primary causes of land degradation; combined, they are responsible for about 84% of the global extent of degraded land, making excessive erosion one of the most significant environmental problems world-wide.

Intensive agriculture, deforestation, roads, anthropogenic climate change and urban sprawl are amongst the most significant human activities in regard to their effect on stimulating erosion. However, there are many prevention and remediation practices that can curtail or limit erosion of vulnerable soils.

A natural arch produced by the wind erosion of differentially weathered rock in Jebel Kharaz, Jordan.

A wave-like sea cliff produced by coastal erosion, in Jinshitan Coastal National Geopark, Dalian, Liaoning Province, China.

Physical Processes

Rainfall and Surface Runoff

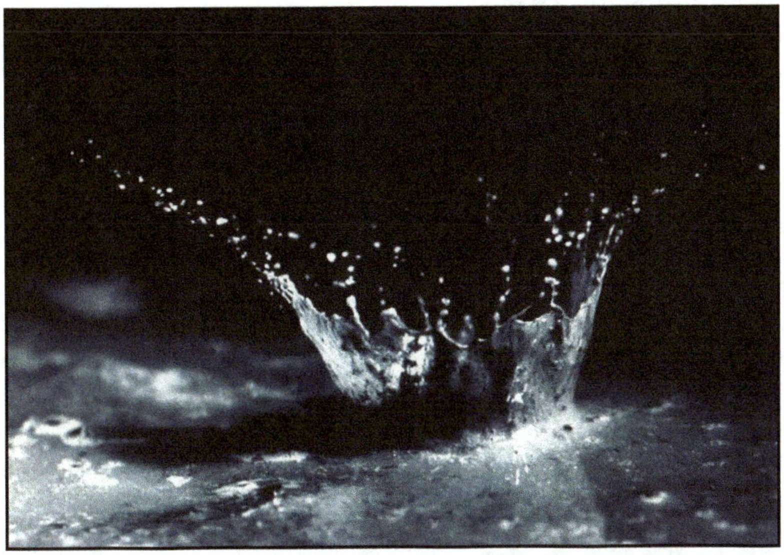

Soil and water being splashed by the impact of a single raindrop.

Rainfall, and the surface runoff which may result from rainfall, produces four main types of soil erosion: *splash erosion*, *sheet erosion*, *rill erosion*, and *gully erosion*. Splash erosion is generally seen as the first and least severe stage in the soil erosion process, which is followed by sheet erosion, then rill erosion and finally gully erosion (the most severe of the four).

In *splash erosion*, the impact of a falling raindrop creates a small crater in the soil, ejecting soil particles. The distance these soil particles travel can be as much as 0.6 m (two feet) vertically and 1.5 m (five feet) horizontally on level ground.

If the soil is saturated, or if the rainfall rate is greater than the rate at which water can infiltrate into the soil, surface runoff occurs. If the runoff has sufficient flow energy, it will transport loosened soil particles (sediment) down the slope. *Sheet erosion* is the transport of loosened soil particles by overland flow.

A spoil tip covered in rills and gullies due to erosion processes caused by rainfall: Rummu, Estonia.

Rill erosion refers to the development of small, ephemeral concentrated flow paths which function as both sediment source and sediment delivery systems for erosion on hillslopes. Generally, where water erosion rates on disturbed upland areas are greatest, rills are active. Flow depths in rills are typically of the order of a few centimetres (about an inch) or less and along-channel slopes may be quite steep. This means that rills exhibit hydraulic physics very different from water flowing through the deeper, wider channels of streams and rivers.

Gully erosion occurs when runoff water accumulates and rapidly flows in narrow channels during or immediately after heavy rains or melting snow, removing soil to a considerable depth.

Rivers and Streams

Dobbingstone Burn, Scotland, showing two different types of erosion affecting the same place. Valley erosion is occurring due to the flow of the stream, and the boulders and stones (and much of the soil) that are lying on the stream's banks are glacial till that was left behind as ice age glaciers flowed over the terrain.

Valley or *stream erosion* occurs with continued water flow along a linear feature. The erosion is both downward, deepening the valley, and headward, extending the valley into the hillside, creating head cuts and steep banks. In the earliest stage of stream erosion, the erosive activity is dominantly vertical, the valleys have a typical V cross-section and the stream gradient is relatively steep. When some base level is reached, the erosive activity switches to lateral erosion, which widens the valley floor and creates a narrow floodplain. The stream gradient becomes nearly flat, and lateral deposition of sediments becomes important as the stream meanders across the valley floor. In all stages of stream erosion, by far the most erosion occurs during times of flood, when more and faster-moving water is available to carry a larger sediment load. In such processes, it is not the water alone that erodes: suspended abrasive particles, pebbles and boulders can also act erosively as they traverse a surface, in a process known as *traction*.

Bank erosion is the wearing away of the banks of a stream or river. This is distinguished from changes on the bed of the watercourse, which is referred to as *scour*. Erosion and changes in the form of river banks may be measured by inserting metal rods into the bank and marking the position of the bank surface along the rods at different times.

Thermal erosion is the result of melting and weakening permafrost due to moving water. It can occur both along rivers and at the coast. Rapid river channel migration observed in the Lena River of Siberia is due to thermal erosion, as these portions of the banks are composed of permafrost-cemented non-cohesive materials. Much of this erosion occurs as the weakened banks fail in large slumps. Thermal erosion also affects the Arctic coast, where wave action and near-shore temperatures combine to undercut permafrost bluffs along the shoreline and cause them to fail. Annual erosion rates along a 100-kilometre (62-mile) segment of the Beaufort Sea shoreline averaged 5.6 metres (18 feet) per year from 1955 to 2002.

Coastal Erosion

Wave cut platform caused by erosion of cliffs by the sea, at Southerndown in South Wales.

Erosion of the boulder clay (of Pleistocene age) along cliffs of Filey Bay, Yorkshire, England.

Shoreline erosion, which occurs on both exposed and sheltered coasts, primarily occurs through the action of currents and waves but sea level (tidal) change can also play a role.

Hydraulic action takes place when air in a joint is suddenly compressed by a wave closing the entrance of the joint. This then cracks it. *Wave pounding* is when the sheer energy of the wave hitting the cliff or rock breaks pieces off. *Abrasion* or *corrasion* is caused by waves launching seaload at the cliff. It is the most effective and rapid form of shoreline erosion. *Corrosion* is the dissolving of rock by carbonic acid in sea water. Limestone cliffs are particularly vulnerable to this kind of erosion. *Attrition* is where particles/seaload carried by the waves are worn down as they hit each other and the cliffs. This then makes the material easier to wash away. The material ends up as shingle and sand. Another significant source of erosion, particularly on carbonate coastlines, is the boring, scraping and grinding of organisms, a process termed *bioerosion*.

Sediment is transported along the coast in the direction of the prevailing current (longshore drift). When the upcurrent amount of sediment is less than the amount being carried away, erosion occurs. When the upcurrent amount of sediment is greater, sand or gravel banks will tend to form as a result of deposition. These banks may slowly migrate along the coast in the direction of the longshore drift, alternately protecting and exposing parts of the coastline. Where there is a bend in the coastline, quite often a buildup of eroded material occurs forming a long narrow bank (a spit). Armoured beaches and submerged offshore sandbanks may also protect parts of a coastline from erosion. Over the years, as the shoals gradually shift, the erosion may be redirected to attack different parts of the shore.

Chemical Erosion

Chemical erosion is the loss of matter in a landscape in the form of solutes. Chemical erosion is usually calculated from the solutes found in streams. Anders Rapp pioneered the study of chemical erosion in his work about Kärkevagge published in 1960.

Glaciers

Glacial moraines above Lake Louise, in Alberta, Canada.

Glaciers erode predominantly by three different processes: abrasion/scouring, plucking, and ice thrusting. In an abrasion process, debris in the basal ice scrapes along the bed, polishing and gouging the underlying rocks, similar to sandpaper on wood. Glaciers can also cause pieces of bedrock to crack off in the process of plucking. In ice thrusting, the glacier freezes to its bed, then as it surges forward, it moves large sheets of frozen sediment at the base along with the glacier. This method produced some of the many thousands of lake basins that dot the edge of the Canadian Shield. The erosion caused by glaciers worldwide erodes mountains so effectively that the term *glacial buzz-saw* has become widely used, which describes the limiting effect of glaciers on the height of mountain ranges. As mountains grow higher, they generally allow for more glacial activity (especially in the accumulation zone above the glacial equilibrium line altitude), which causes increased rates of erosion of the mountain, decreasing mass faster than isostatic rebound can add to the mountain. This provides a good example of a negative feedback loop. Ongoing research is showing that while glaciers tend to decrease mountain size, in some areas, glaciers can actually reduce the rate of erosion, acting as a *glacial armour*.

These processes, combined with erosion and transport by the water network beneath the glacier, leave moraines, drumlins, ground moraine (till), kames, kame deltas, moulins, and glacial erratics in their wake, typically at the terminus or during glacier retreat.

Floods

At extremely high flows, kolks, or vortices are formed by large volumes of rapidly rushing water. Kolks cause extreme local erosion, plucking bedrock and creating pothole-type geographical features called Rock-cut basins. Examples can be seen in the flood regions result from glacial Lake Missoula, which created the channeled scablands in the Columbia Basin region of eastern Washington.

Wind Erosion

Árbol de Piedra, a rock formation in the Altiplano, Bolivia sculpted by wind erosion.

Wind erosion is a major geomorphological force, especially in arid and semi-arid regions. It is also a major source of land degradation, evaporation, desertification, harmful airborne dust, and crop damage—especially after being increased far above natural rates by human activities such as deforestation, urbanization, and agriculture.

Wind erosion is of two primary varieties: *deflation*, where the wind picks up and carries away loose particles; and *abrasion*, where surfaces are worn down as they are struck by airborne particles carried by wind. Deflation is divided into three categories: (1) *surface creep*, where larger, heavier particles slide or roll along the ground; (2) *saltation*, where particles are lifted a short height into the air, and bounce and saltate across the surface of the soil; and (3) *suspension*, where very small and light particles are lifted into the air by the wind, and are often carried for long distances. Saltation is responsible for the majority (50-70%) of wind erosion, followed by suspension (30-40%), and then surface creep (5-25%).

Wind erosion is much more severe in arid areas and during times of drought. For example, in the Great Plains, it is estimated that soil loss due to wind erosion can be as much as 6100 times greater in drought years than in wet years.

Mass Movement

Wadi in Makhtesh Ramon, Israel, showing gravity collapse erosion on its banks.

Mass movement is the downward and outward movement of rock and sediments on a sloped surface, mainly due to the force of gravity.

Mass movement is an important part of the erosional process, and is often the first stage in the breakdown and transport of weathered materials in mountainous areas. It moves material from higher elevations to lower elevations where other eroding agents such as streams and glaciers can then pick up the material and move it to even lower elevations. Mass-movement processes are always occurring continuously on all slopes; some mass-movement processes act very slowly; others occur very suddenly, often with disastrous results. Any perceptible down-slope movement of rock or sediment is often referred to in general terms as a landslide. However, landslides can be classified in a much more detailed way that reflects the mechanisms responsible for the movement and the velocity at which the movement occurs. One of the visible topographical manifestations of a very slow form of such activity is a scree slope.

Slumping happens on steep hillsides, occurring along distinct fracture zones, often within materials like clay that, once released, may move quite rapidly downhill. They will often show a spoon-shaped isostatic depression, in which the material has begun to slide downhill. In some cases, the slump is caused by water beneath the slope weakening it. In many cases it is simply the result of poor engineering along highways where it is a regular occurrence.

Surface creep is the slow movement of soil and rock debris by gravity which is usually not perceptible except through extended observation. However, the term can also describe the rolling of dislodged soil particles 0.5 to 1.0 mm (0.02 to 0.04 in) in diameter by wind along the soil surface.

Factors Affecting Erosion Rates

Climate

The amount and intensity of precipitation is the main climatic factor governing soil erosion by water. The relationship is particularly strong if heavy rainfall occurs at times when, or in locations where, the soil's surface is not well protected by vegetation. This might be during periods when agricultural activities leave the soil bare, or in semi-arid regions where vegetation is naturally sparse. Wind erosion requires strong winds, particularly during times of drought when vegetation is sparse and soil is dry (and so is more erodible). Other climatic factors such as average temperature and temperature range may also affect erosion, via their effects on vegetation and soil properties. In general, given similar vegetation and ecosystems, areas with more precipitation (especially high-intensity rainfall), more wind, or more storms are expected to have more erosion.

In some areas of the world (e.g. the mid-western USA), rainfall intensity is the primary determinant of erosivity, with higher intensity rainfall generally resulting in more soil erosion by water. The size and velocity of rain drops is also an important factor. Larger and higher-velocity rain drops have greater kinetic energy, and thus their impact will displace soil particles by larger distances than smaller, slower-moving rain drops.

In other regions of the world (e.g. western Europe), runoff and erosion result from relatively low intensities of stratiform rainfall falling onto previously saturated soil. In such situations, rainfall amount rather than intensity is the main factor determining the severity of soil erosion by water.

Vegetative Cover

Vegetation acts as an interface between the atmosphere and the soil. It increases the permeability of the soil to rainwater, thus decreasing runoff. It shelters the soil from winds, which results in decreased wind erosion, as well as advantageous changes in microclimate. The roots of the plants bind the soil together, and interweave with other roots, forming a more solid mass that is less susceptible to both water and wind erosion. The removal of vegetation increases the rate of surface erosion.

Topography

The topography of the land determines the velocity at which surface runoff will flow, which in turn determines the erosivity of the runoff. Longer, steeper slopes (especially those without adequate vegetative cover) are more susceptible to very high rates of erosion during heavy rains than shorter, less steep slopes. Steeper terrain is also more prone to mudslides, landslides, and other forms of gravitational erosion processes.

Tectonics

Tectonic processes control rates and distributions of erosion at the Earth's surface. If tectonic action causes part of the Earth's surface (e.g., a mountain range) to be raised or lowered relative to surrounding areas, this must necessarily change the gradient of the land surface. Because erosion rates are almost always sensitive to local slope, this will change the rates of erosion in the uplifted area. Active tectonics also brings fresh, unweathered rock towards the surface, where it is exposed to the action of erosion.

However, erosion can also affect tectonic processes. The removal by erosion of large amounts of rock from a particular region, and its deposition elsewhere, can result in a lightening of the load on the lower crust and mantle. Because tectonic processes are driven by gradients in the stress field developed in the crust, this unloading can in turn cause tectonic or isostatic uplift in the region. In some cases, it has been hypothesised that these twin feedbacks can act to localise and enhance zones of very rapid exhumation of deep crustal rocks beneath places on the Earth's surface with extremely high erosion rates, for example, beneath the extremely steep terrain of Nanga Parbat in the western Himalayas. Such a place has been called a "tectonic aneurysm".

Erosion of Earth Systems

Mountain Ranges

Mountain ranges are known to take many million of years to erode to the degree they effectively cease to exist. Scholars Pitman and Golovchenko estimate that it takes probably more than 450 million years to erode a mountain mass similar to the Himalaya into an almost-flat peneplain if there are no major sea-level changes. Erosion of mountains massifs can create a pattern of equally high summits called summit accordance.

Examples of heavily eroded mountain ranges include the Timanides of Northern Russia. Erosion of this orogen has produced sediments that are now found in the East European Platform, including the Cambrian Sablya Formation near Lake Ladoga. Studies of these sediments points that its likely that the erosion of the orogen was beginning in the Cambrian and then became stronger in Ordovician.

Soils

If the rate of erosion is higher than the rate of soil formation the soils are being destroyed by erosion. Where soil is not destroyed by erosion, erosion can in some cases prevent the formation of soil features that form slowly. Inceptisols are common soils that form in areas of fast erosion.

While erosion of soils is a natural process, human activities have increased by 10-40 times the rate at which erosion is occurring globally. Excessive (or accelerated) erosion causes both "on-site" and "off-site" problems. On-site impacts include decreases in agricultural productivity and (on natural landscapes) ecological collapse, both because of loss of the nutrient-rich upper soil layers. In some cases, the eventual end result is desertification. Off-site effects include sedimentation of waterways and eutrophication of water bodies, as well as sediment-related damage to roads and houses. Water and wind erosion are the two primary causes of land degradation; combined, they are responsible for about 84% of the global extent of degraded land, making excessive erosion one of the most significant environmental problems world-wide.

Prevention of Soil Loss

Soil Conservation

Soil conservation is the prevention of soil loss from erosion or reduced fertility caused by over usage, acidification, salinization or other chemical soil contamination.

Slash-and-burn and other unsustainable methods of subsistence farming are practiced in some lesser developed areas. A sequel to the deforestation is typically large scale erosion, loss of soil nutrients and sometimes total desertification.

Techniques for improved soil conservation include crop rotation, cover crops, conservation tillage and planted windbreaks and affect both erosion and fertility. When plants, especially trees, die, they decay and become part of the soil. Code 330 defines standard methods recommended by the US Natural Resources Conservation Service.

Erosion barriers on disturbed slope, Marin County, California.

Contour plowing, Pennsylvania 1938. The rows formed slow water run-off during rainstorms to prevent soil erosion and allows the water time to infiltrate into the soil.

Farmers have practiced soil conservation for millennia. Conservation practices fall in multiple categories:

Contour Plowing

Contour plowing orients crop furrows following the contour lines of the farmed area. Furrows move left and right to maintain a constant altitude, which reduces runoff. Contour plowing was practiced by the ancient Phoenicians, and is effective for slopes between two and ten percent. Contour plowing can increase crop yields from 10 to 50 percent, partially as a result from greater soil retention.

Terracing or Terrace Farming

Terracing is the practice of creating nearly level areas in a hillside area. The terraces form a series of steps, each at a higher level than the previous. Terraces are protected from erosion by other soil barriers.

Terraced farming is more common on small farms and in underdeveloped countries, since mechanized equipment is difficult to deploy in this setting.

Keyline Design

Keyline design is an enhancement of contour farming, where the total watershed properties are taken into account in forming the contour lines.

Perimeter Runoff Control

Tree, shrubs and ground-cover are effective perimeter treatment for soil erosion prevention, by impeding surface flows. A special form of this perimeter or inter-row treatment is the use of a "grass way" that both channels and dissipates runoff through surface friction, impeding surface runoff and encouraging infiltration of the slowed surface water.

Windbreaks

Windbreaks are sufficiently dense rows of trees at the windward exposure of an agricultural field subject to wind erosion. Evergreen species provide year-round protection; however, as long as foliage is present in the seasons of bare soil surfaces, the effect of deciduous trees may be adequate.

Cover Crops/Crop Rotation

Cover crops such as legumes, white turnip, radishes and other species are rotated with cash crops to blanket the soil year-round and act as green manure that replenishes nitrogen and other critical nutrients. Cover crops also help suppress weeds.

Soil-Conservation Farming

Soil-conservation farming involves no-till farming, "green manures" and other soil-enhancing practices. Such farming methods attempt to mimic the biology of virgin land. They can revive damaged soil, minimize erosion, encourage plant growth, eliminate the use of nitrogen fertilizer or fungicide, produce above-average yields and protect crops during droughts or flooding. The result is less labor and lower costs that increase farmers' profits. No-till farming and cover crops act as sinks for nitrogen and other nutrients. This increases the amount of soil organic matter.

Repeated plowing/tilling degrades soil, killing its beneficial fungi and earthworms. Once damaged, soil may take multiple seasons to fully recover, even in optimal circumstances.

Critics argue that no-till and related methods are impractical and too expensive for many growers, partly because it requires new equipment. They cite advantages for conventional tilling depending on the geography, crops and soil conditions. Some farmers claimed that no-till complicates weed control, delays planting and that post-harvest residues, especially for corn, are hard to manage.

Salinity Management

Salt deposits on the former bed of the Aral Sea.

Salinity in soil is caused by irrigating with salty water. Water then evaporates from the soil leaving the salt behind. Salt breaks down the soil structure, causing infertility and reduced growth.

The ions responsible for salination are: sodium (Na^{+}), potassium (K^{+}), calcium (Ca^{2+}), magnesium (Mg^{2+}) and chlorine (Cl^{-}). Salinity is estimated to affect about one third of the earth's arable land. Soil salinity adversely affects crop metabolism and erosion usually follows.

Salinity occurs on drylands from overirrigation and in areas with shallow saline water tables. Over-irrigation deposits salts in upper soil layers as a byproduct of soil infiltration; irrigation merely increases the rate of salt deposition. The best-known case of shallow saline water table capillary action occurred in Egypt after the 1970 construction of the Aswan Dam. The change in the groundwater level led to high salt concentrations in the water table. The continuous high level of the water table led to soil salination.

Use of humic acids may prevent excess salination, especially given excessive irrigation. Humic acids can fix both anions and cations and eliminate them from root zones.

Planting species that can tolerate saline conditions to produce surface cover can preserve soil salinity can be reduced. Salt-tolerant plants include saltbush, a plant found in much of North America and in the Mediterranean regions of Europe.

Soil Organisms

Yellow fungus, a mushroom that assists in organic decay.

When worms excrete egesta in the form of casts, a balanced selection of minerals and plant nutrients is made into a form accessible for root uptake. Earthworm casts are five times richer in available nitrogen, seven times richer in available phosphates and eleven times richer in available potash than the surrounding upper 150 millimetres (5.9 in) of soil. The weight of casts produced may be greater than 4.5 kg per worm per year. By burrowing, the earthworm improves soil porosity, creating channels that enhance the processes of aeration and drainage.

Other important soil organisms include nematodes, mycorrizha and bacteria.

Degraded soil requires synthetic fertilizer to produce high yields. Lacking structure increases erosion and carries nitrogen and other pollutants into rivers and streams.

Each one percent increase in soil organic matter helps soil hold 20,000 gallons more water per acre.

Mineralization

To allow plants full realization of their phytonutrient potential, active mineralization of the soil is sometimes undertaken. This can involve adding crushed rock or chemical soil supplements. In either case the purpose is to combat mineral depletion. A broad range of minerals can be used, including common substances such as phosphorus and more exotic substances such as zinc and selenium. Extensive research examines the phase transitions of minerals in soil with aqueous contact.

Flooding can bring significant sediment to an alluvial plain. While this effect may not be desirable if floods endanger life or if the sediment originates from productive land, this process of addition to a floodplain is a natural process that can rejuvenate soil chemistry through mineralization and macronutrient addition.

US

For some crops, no-tillage acreage nearly doubled from 2000 to 2015. For soybeans, for example, it rose to 30 million acres in 2012 from 16.5 million acres in 1996. Use of cover crops was rising in acreage about 30 percent a year, according to surveys, though the total remains small.

Soil Liquefaction

Some effects of liquefaction after the 1964 Niigata earthquake.

Liquefaction allowed this sewer to float upward – 2004 Chūetsu earthquake.

The effect of liquefaction in Christchurch, New Zealand, during the Mw 6.3 February 2011 Christchurch earthquake.

Soil liquefaction describes a phenomenon whereby a saturated or partially saturated soil substantially loses strength and stiffness in response to an applied stress, usually earthquake shaking or other sudden change in stress condition, causing it to behave like a liquid.

In soil mechanics the term "liquefied" was first used by Allen Hazen in reference to the 1918 failure of the Calaveras Dam in California. He described the mechanism of flow liquefaction of the embankment dam as follows:

If the pressure of the water in the pores is great enough to carry all the load, it will have the effect of holding the particles apart and of producing a condition that is practically equivalent to that of quicksand... the initial movement of some part of the material might result in accumulating pressure, first on one point, and then on another, successively, as the early points of concentration were liquefied.

The phenomenon is most often observed in saturated, loose (low density or uncompacted), sandy soils. This is because a loose sand has a tendency to compress when a load is applied; dense sands by contrast tend to expand in volume or 'dilate'. If the soil is saturated by water, a condition that often exists when the soil is below the ground water table or sea level, then water fills the gaps between soil grains ('pore spaces'). In response to the soil compressing, this water increases in pressure and attempts to flow out from the soil to zones of low pressure (usually upward towards the ground surface). However, if the loading is rapidly applied and large enough, or is repeated many times (e.g. earthquake shaking, storm wave loading) such that it does not flow out in time before the next cycle of load is applied, the water pressures may build to an extent where they exceed the contact stresses between the grains of soil that keep them in contact with each other. These contacts between grains are the means by which the weight from buildings and overlying soil layers are transferred from the ground surface to layers of soil or rock at greater depths. This loss of soil structure causes it to lose all of its strength (the ability to transfer shear stress) and it may be observed to flow like a liquid (hence 'liquefaction').

Although the effects of liquefaction have been long understood, it was more thoroughly brought to the attention of engineers after the 1964 Niigata earthquake and 1964 Alaska earthquake. It was also a major factor in the destruction in San Francisco's Marina District during the 1989 Loma Prieta earthquake, and in Port of Kobe during the 1995 Great Hanshin earthquake. More recently liquefaction was largely responsible for extensive damage to residential properties in the eastern suburbs and satellite townships of Christchurch, New Zealand during the 2010 Canterbury earthquake and more extensively again following the Christchurch earthquakes that followed in early and mid-2011.

The building codes in many developed countries require engineers to consider the effects of soil liquefaction in the design of new buildings and infrastructure such as bridges, embankment dams and retaining structures.

Technical Definitions

A state of 'soil liquefaction' occurs when the effective stress of soil is reduced to essentially zero, which corresponds to a complete loss of shear strength. This may be initiated by either monotonic loading (e.g. single sudden occurrence of a change in stress – examples include an increase in load on an embankment or sudden loss of toe support) or cyclic loading (e.g. repeated change in stress condition – examples include wave loading or earthquake shaking). In both cases a soil in a saturated loose state, and one which may generate significant pore water pressure on a change in load are the most likely to liquefy. This is because a loose soil has the tendency to compress when sheared, generating large excess porewater pressure as load is transferred from the soil skeleton to adjacent pore water during undrained loading. As pore water pressure rises a progressive loss of strength of the soil occurs as effective stress is reduced. It is more likely to occur in sandy or non-plastic silty soils, but may in rare cases occur in gravels and clays.

A 'flow failure' may initiate if the strength of the soil is reduced below the stresses required to maintain equilibrium of a slope or footing of a building for instance. This can occur due to monotonic loading or cyclic loading, and can be sudden and catastrophic. A historical example is the Aberfan disaster. Casagrande referred to this type of phenomena as 'flow liquefaction' although a state of zero effective stress is not required for this to occur.

The term 'cyclic liquefaction' refers to the occurrence of a state of soil when large shear strains have accumulated in response to cyclic loading. A typical reference strain for the approximate occurrence of zero effective stress is 5% double amplitude shear strain. This is a soil test based definition, usually performed via cyclic triaxial, cyclic direct simple shear, or cyclic torsional shear type apparatus. These tests are performed to determine a soil's resistance to liquefaction by observing the number of cycles of loading at a particular shear stress amplitude before it 'fails'. Failure here is defined by the aforementioned shear strain criteria.

The term 'cyclic mobility' refers to the mechanism of progressive reduction of effective stress due to cyclic loading. This may occur in all soil types including dense soils. However, on reaching a state of zero effective stress such soils immediate dilate and regain strength. Thus shear strains are significantly less than a true state of soil liquefaction whereby a loose soil exhibits flow type phenomena.

Occurrence

Liquefaction is more likely to occur in loose to moderately saturated granular soils with poor drainage, such as silty sands or sands and gravels capped or containing seams of impermeable sediments. During wave loading, usually cyclic undrained loading, e.g. seismic loading, loose sands tend to decrease in volume, which produces an increase in their pore water pressures and consequently a decrease in shear strength, i.e. reduction in effective stress.

Deposits most susceptible to liquefaction are young (Holocene-age, deposited within the last 10,000 years) sands and silts of similar grain size (well-sorted), in beds at least metres thick, and saturated with water. Such deposits are often found along stream beds, beaches, dunes, and areas where windblown silt (loess) and sand have accumulated. Some examples of soil liquefaction include quicksand, quick clay, turbidity currents, and earthquake induced liquefaction.

Depending on the initial void ratio, the soil material can respond to loading either *strain-softening* or *strain-hardening*. Strain-softened soils, e.g. loose sands, can be triggered to collapse, either monotonically or cyclically, if the static shear stress is greater than the ultimate or steady-state shear strength of the soil. In this case *flow liquefaction* occurs, where the soil deforms at a low constant residual shear stress. If the soil strain-hardens, e.g. moderately dense to dense sand, flow liquefaction will generally not occur. However, *cyclic softening* can occur due to cyclic undrained loading, e.g. earthquake loading. Deformation during cyclic loading will depend on the density of the soil, the magnitude and duration of the cyclic loading, and amount of shear stress reversal. If stress reversal occurs, the effective shear stress could reach zero, then *cyclic liquefaction* can take place. If stress reversal does not occur, zero effective stress is not possible to occur, then *cyclic mobility* takes place.

The resistance of the cohesionless soil to liquefaction will depend on the density of the soil, confining stresses, soil structure (fabric, age and cementation), the magnitude and duration of the cyclic loading, and the extent to which shear stress reversal occurs.

Earthquake Liquefaction

Sand boils that erupted during the 2011 Christchurch earthquake.

The pressures generated during large earthquakes with many cycles of shaking can cause the liquefied sand and excess water to force its way to the ground surface from several metres below the ground. This is often observed as "sand boils" also called "sand blows" or "sand volcanoes" (as they appear to form small volcanic craters) at the ground surface. The phenomenon may incorporate both flow of already liquefied sand from a layer below ground, and a quicksand effect whereby upward flow of water initiates liquefaction in overlying non-liquefied sandy deposits due to buoyancy.

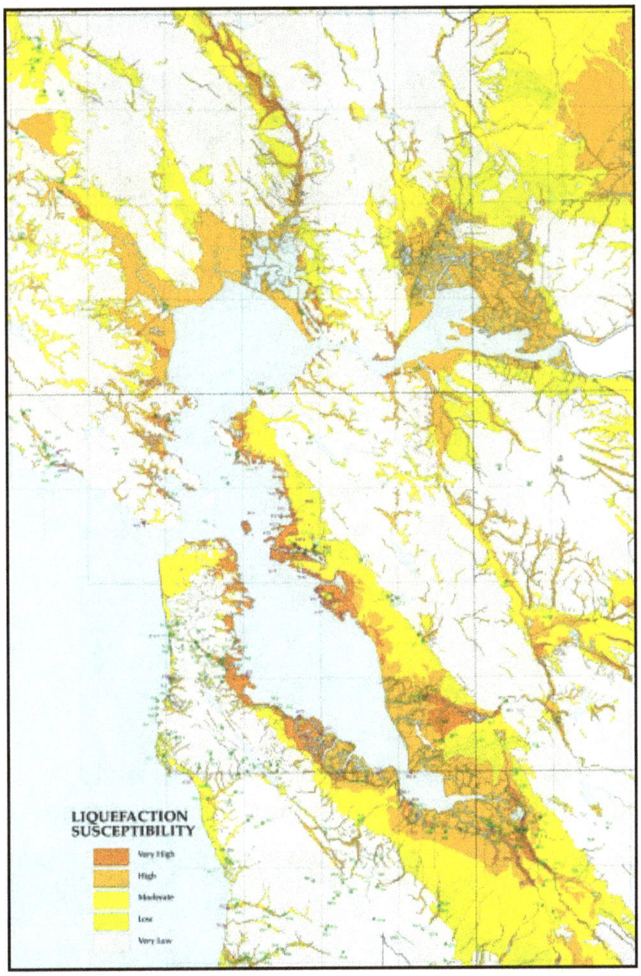

A liquefaction susceptibility map – excerpt of USGS map for the San Francisco Bay Area. Many areas of concern in this region are also densely urbanized.

The other common observation is land instability – cracking and movement of the ground down slope or towards unsupported margins of rivers, streams, or the coast. The failure of ground in this manner is called 'lateral spreading', and may occur on very shallow slopes of angles of only 1 or 2 degrees from the horizontal.

One positive aspect of soil liquefaction is the tendency for the effects of earthquake shaking to be significantly damped (reduced) for the remainder of the earthquake. This is because liquids do not support a shear stress and so once the soil liquefies due to shaking, subsequent earthquake shaking (transferred through ground by shear waves) is not transferred to buildings at the ground surface.

Studies of liquefaction features left by prehistoric earthquakes, called paleoliquefaction or paleoseismology, can reveal a great deal of information about earthquakes that occurred before records were kept or accurate measurements could be taken.

Soil liquefaction induced by earthquake shaking is also a major contributor to urban seismic risk.

Effects

The effects of lateral spreading (River Road in Christchurch following the 2011 Christchurch earthquake).

Damage in Brooklands from the 2010 Canterbury earthquake, where buoyancy caused by soil liquefaction pushed up an underground service including this manhole.

The effects of soil liquefaction on the built environment can be extremely damaging. Buildings whose foundations bear directly on sand which liquefies will experience a sudden loss of support, which will result in drastic and irregular settlement of the building causing structural damage, including cracking of foundations and damage to the building structure itself, or may leave the structure unserviceable afterwards, even without structural damage. Where a thin crust of non-liquefied soil exists between building foundation and liquefied soil, a 'punching shear' type foundation failure may occur. The irregular settlement of ground may also break underground utility lines. The upward pressure applied by the movement of liquefied soil through the crust layer can crack weak foundation slabs and enter buildings through service ducts, and may allow water to damage the building contents and electrical services.

Bridges and large buildings constructed on pile foundations may lose support from the adjacent soil and buckle, or come to rest at a tilt after shaking.

Sloping ground and ground next to rivers and lakes may slide on a liquefied soil layer (termed 'lateral spreading'), opening large cracks or fissures in the ground, and can cause significant damage to buildings, bridges, roads and services such as water, natural gas, sewerage, power and telecommunications installed in the affected ground. Buried tanks and manholes may float in the liquefied soil due to buoyancy. Earth embankments such as flood levees and earth dams may lose stability or collapse if the material comprising the embankment or its foundation liquefies.

Mitigation Methods

Methods to mitigate the effects of soil liquefaction have been devised by earthquake engineers and include various soil compaction techniques such as vibro compaction (compaction of the soil by depth vibrators), dynamic compaction, and vibro stone columns. These methods result in the densification of soil and enable buildings to withstand soil liquefaction.

Existing buildings can be mitigated by injecting grout into the soil to stabilize the layer of soil that is subject to liquefaction.

Quicksand

Quicksand forms when water saturates an area of loose sand and the ordinary sand is agitated. When the water trapped in the batch of sand cannot escape, it creates liquefied soil that can no longer support weight. Quicksand can be formed by standing or (upwards) flowing underground water (as from an underground spring), or by earthquakes. In the case of flowing underground water, the force of the water flow opposes the force of gravity, causing the granules of sand to be more buoyant. In the case of earthquakes, the shaking force can increase the pressure of shallow groundwater, liquefying sand and silt deposits. In both cases, the liquefied surface loses strength, causing buildings or other objects on that surface to sink or fall over.

The saturated sediment may appear quite solid until a change in pressure or shock initiates the liquefaction, causing the sand to form a suspension with each grain surrounded by a thin film of water. This cushioning gives quicksand, and other liquefied sediments, a spongy, fluidlike texture. Objects in the liquefied sand sink to the level at which the weight of the object is equal to the weight of the displaced sand/water mix and the object *floats* due to its buoyancy.

Quick clay

Quick clay, also known as *Leda Clay* in Canada, is a water-saturated gel, which in its solid form resemble a unique form of highly sensitive clay. This clay has a tendency to change from a relatively stiff condition to a liquid mass when it is disturbed. This gradual change in appearance from solid to liquid is a process known as spontaneous liquefaction. The clay retains a solid structure despite the high water content (up to 80% by volume), because surface tension holds water-coated flakes of clay together in a delicate structure. When the structure is broken by a shock or sufficient shear, it turns to a fluid state.

Quick clay is only found in the northern countries such as Russia, Canada, Alaska in the U.S., Norway, Sweden, and Finland, which were glaciated during the Pleistocene epoch.

Quick clay has been the underlying cause of many deadly landslides. In Canada alone, it has been associated with more than 250 mapped landslides. Some of these are ancient, and may have been triggered by earthquakes.

Turbidity Currents

Submarine landslides are turbidity currents and consist of water saturated sediments flowing downslope. An example occurred during the 1929 Grand Banks earthquake that struck the continental slope off the coast of Newfoundland. Minutes later, transatlantic telephone cables began breaking sequentially, farther and farther downslope, away from the epicenter. Twelve cables were snapped in a total of 28 places. Exact times and locations were recorded for each break. Investigators suggested that a 60-mile-per-hour (100 km/h) submarine *landslide* or turbidity current of water saturated sediments swept 400 miles (600 km) down the continental slope from the earthquake's epicenter, snapping the cables as it passed.

Soil Contamination

Soil contamination or soil pollution as part of land degradation is caused by the presence of xenobiotic (human-made) chemicals or other alteration in the natural soil environment. It is typically caused by industrial activity, agricultural chemicals, or improper disposal of waste. The most common chemicals involved are petroleum hydrocarbons, polynuclear aromatic hydrocarbons (such as naphthalene and benzo(a) pyrene), solvents, pesticides, lead, and other heavy metals. Contamination is correlated with the degree of industrialization and intensity of chemical usage.

The concern over soil contamination stems primarily from health risks, from direct contact with the contaminated soil, vapors from the contaminants, and from secondary contamination of water supplies within and underlying the soil. Mapping of contaminated soil sites and the resulting cleanup are time consuming and expensive tasks, requiring extensive amounts of geology, hydrology, chemistry, computer modeling skills, and GIS in Environmental Contamination, as well as an appreciation of the history of industrial chemistry.

Excavation showing soil contamination at a disused gasworks.

In North America and Western Europe the extent of contaminated land is best known, with many of countries in these areas having a legal framework to identify and deal with this environmental problem. Developing countries tend to be less tightly regulated despite some of them having undergone significant industrialization.

Causes

Soil pollution can be caused by the following (non-exhaustive list!):

- Oil drilling.
- Mining and activities by other heavy industries.
- Accidental spills as may happen during deforestation activities, etc.
- Corrosion of underground storage tanks (including piping used to transmit the contents).
- Acid rain (in turn caused by air pollution).
- Intensive farming.
- Agrochemicals, such as pesticides, herbicides and fertilizers.
- Industrial accidents.

- Road debris.
- Drainage of contaminated surface water into the soil.
- Waste disposal:
  - Oil and fuel dumping.
  - Nuclear wastes.
  - Direct discharge of industrial wastes to the soil.
  - Landfill and illegal dumping.
  - Coal ash.
  - Electronic waste.
  - Ammunitions and agents of war.

The most common chemicals involved are petroleum hydrocarbons, solvents, pesticides, lead, and other heavy metals.

In a wider sense, genetically modified plants (GMP) can count as a risk factor for soils, because of their potential to affect the soil fauna. Any activity that leads to other forms of soil degradation (erosion, compaction, etc.) may indirectly worsen the contamination effects in that soil remediation becomes more tedious.

Historical deposition of coal ash used for residential, commercial, and industrial heating, as well as for industrial processes such as ore smelting, were a common source of contamination in areas that were industrialized before about 1960. Coal naturally concentrates lead and zinc during its formation, as well as other heavy metals to a lesser degree. When the coal is burned, most of these metals become concentrated in the ash (the principal exception being mercury). Coal ash and slag may contain sufficient lead to qualify as a "characteristic hazardous waste", defined in the USA as containing more than 5 mg/l of extractable lead using the TCLP procedure. In addition to lead, coal ash typically contains variable but significant concentrations of polynuclear aromatic hydrocarbons (PAHs; e.g., benzo(a)anthracene, benzo(b)fluoranthene, benzo(k)fluoranthene, benzo(a)pyrene, indeno(cd)pyrene, phenanthrene, anthracene, and others). These PAHs are known human carcinogens and the acceptable concentrations of them in soil are typically around 1 mg/kg. Coal ash and slag can be recognised by the presence of off-white grains in soil, gray heterogeneous soil, or (coal slag) bubbly, vesicular pebble-sized grains.

Treated sewage sludge, known in the industry as biosolids, has become controversial as a "fertilizer". As it is the byproduct of sewage treatment, it generally contains more contaminants such as organisms, pesticides, and heavy metals than other soil.

In the European Union, the Urban Waste Water Treatment Directive allows sewage sludge to be sprayed onto land. The volume is expected to double to 185,000 tons of dry solids in 2005. This has good agricultural properties due to the high nitrogen and phosphate content. In 1990/1991, 13% wet weight was sprayed onto 0.13% of the land; however, this is expected to rise 15 fold by 2005. Advocates say there is a need to control this so that pathogenic microorganisms do not get into water courses and to ensure that there is no accumulation of heavy metals in the top soil.

Pesticides and Herbicides

A pesticide is a substance or mixture of substances used to kill a pest. A pesticide may be a chemical substance, biological agent (such as a virus or bacteria), antimicrobial, disinfectant or device used against any pest. Pests include insects, plant pathogens, weeds, mollusks, birds, mammals, fish, nematodes (roundworms) and microbes that compete with humans for food, destroy property, spread or are a vector for disease or cause a nuisance. Although there are benefits to the use of pesticides, there are also drawbacks, such as potential toxicity to humans and other organisms.

Herbicides are used to kill weeds, especially on pavements and railways. They are similar to auxins and most are biodegradable by soil bacteria. However, one group derived from trinitrotoluene (2:4 D and 2:4:5 T) have the impurity dioxin, which is very toxic and causes fatality even in low concentrations. Another herbicide is Paraquat. It is highly toxic but it rapidly degrades in soil due to the action of bacteria and does not kill soil fauna.

Insecticides are used to rid farms of pests which damage crops. The insects damage not only standing crops but also stored ones and in the tropics it is reckoned that one third of the total production is lost during food storage. As with fungicides, the first insecticides used in the nineteenth century were inorganic e.g. Paris Green and other compounds of arsenic. Nicotine has also been used since the late eighteenth century.

There are now two main groups of synthetic insecticides -

1. Organochlorines include DDT, Aldrin, Dieldrin and BHC. They are cheap to produce, potent and persistent. DDT was used on a massive scale from the 1930s, with a peak of 72,000 tonnes used 1970. Then usage fell as the harmful environmental effects were realized. It was found worldwide in fish and birds and was even discovered in the snow in the Antarctic. It is only slightly soluble in water but is very soluble in the bloodstream. It affects the nervous and endocrine systems and causes the eggshells of birds to lack calcium causing them to be easily breakable. It is thought to be responsi-

ble for the decline of the numbers of birds of prey like ospreys and peregrine falcons in the 1950s - they are now recovering. As well as increased concentration via the food chain, it is known to enter via permeable membranes, so fish get it through their gills. As it has low water solubility, it tends to stay at the water surface, so organisms that live there are most affected. DDT found in fish that formed part of the human food chain caused concern, but the levels found in the liver, kidney and brain tissues was less than 1 ppm and in fat was 10 ppm, which was below the level likely to cause harm. However, DDT was banned in the UK and the United States to stop the further buildup of it in the food chain. U.S. manufacturers continued to sell DDT to developing countries, who could not afford the expensive replacement chemicals and who did not have such stringent regulations governing the use of pesticides.

2. Organophosphates, e.g. parathion, methyl parathion and about 40 other insecticides are available nationally. Parathion is highly toxic, methyl-parathion is less so and Malathion is generally considered safe as it has low toxicity and is rapidly broken down in the mammalian liver. This group works by preventing normal nerve transmission as cholinesterase is prevented from breaking down the transmitter substance acetylcholine, resulting in uncontrolled muscle movements.

Agents of War

The disposal of munitions, and a lack of care in manufacture of munitions caused by the urgency of production, can contaminate soil for extended periods. There is little published evidence on this type of contamination largely because of restrictions placed by Governments of many countries on the publication of material related to war effort. However, mustard gas stored during World War II has contaminated some sites for up to 50 years and the testing of Anthrax as a potential biological weapon contaminated the whole island of Gruinard.

Health Effects

Contaminated or polluted soil directly affects human health through direct contact with soil or via inhalation of soil contaminants which have vaporized; potentially greater threats are posed by the infiltration of soil contamination into groundwater aquifers used for human consumption, sometimes in areas apparently far removed from any apparent source of above ground contamination.

Health consequences from exposure to soil contamination vary greatly depending on pollutant type, pathway of attack and vulnerability of the exposed population. Chronic exposure to chromium, lead and other metals, petroleum, solvents, and many pesticide and herbicide formulations can be carcinogenic, can cause congenital disorders, or can cause other chronic health conditions. Industrial or man-made concentrations of naturally occurring substances, such as nitrate and ammonia associated with livestock manure from agricultural operations, have also been identified as health hazards in soil and groundwater.

Chronic exposure to benzene at sufficient concentrations is known to be associated with higher incidence of leukemia. Mercury and cyclodienes are known to induce higher incidences of kidney damage and some irreversible diseases. PCBs and cyclodienes are linked to liver toxicity. Organophosphates and carbomates can induce a chain of responses leading to neuromuscular blockage. Many chlorinated solvents induce liver changes, kidney changes and depression of the central nervous system. There is an entire spectrum of further health effects such as headache, nausea, fatigue, eye irritation and skin rash for the above cited and other chemicals. At sufficient dosages a large number of soil contaminants can cause death by exposure via direct contact, inhalation or ingestion of contaminants in groundwater contaminated through soil.

The Scottish Government has commissioned the Institute of Occupational Medicine to undertake a review of methods to assess risk to human health from contaminated land. The overall aim of the project is to work up guidance that should be useful to Scottish Local Authorities in assessing whether sites represent a significant possibility of significant harm (SPOSH) to human health. It is envisaged that the output of the project will be a short document providing high level guidance on health risk assessment with reference to existing published guidance and methodologies that have been identified as being particularly relevant and helpful. The project will examine how policy guidelines have been developed for determining the acceptability of risks to human health and propose an approach for assessing what constitutes unacceptable risk in line with the criteria for SPOSH as defined in the legislation and the Scottish Statutory Guidance.

Ecosystem Effects

Not unexpectedly, soil contaminants can have significant deleterious consequences for ecosystems. There are radical soil chemistry changes which can arise from the presence of many hazardous chemicals even at low concentration of the contaminant species. These changes can manifest in the alteration of metabolism of endemic microorganisms and arthropods resident in a given soil environment. The result can be virtual eradication of some of the primary food chain, which in turn could have major consequences for predator or consumer species. Even if the chemical effect on lower life forms is small, the lower pyramid levels of the food chain may ingest alien chemicals, which normally become more concentrated for each consuming rung of the food chain. Many of these effects are now well known, such as the concentration of persistent DDT materials for avian consumers, leading to weakening of egg shells, increased chick mortality and potential extinction of species.

Effects occur to agricultural lands which have certain types of soil contamination. Contaminants typically alter plant metabolism, often causing a reduction in crop yields. This has a secondary effect upon soil conservation, since the languishing crops cannot shield the Earth's soil from erosion. Some of these chemical contaminants have long half-lives and in other cases derivative chemicals are formed from decay of primary soil contaminants.

Cleanup Options

Cleanup or environmental remediation is analyzed by environmental scientists who utilize field measurement of soil chemicals and also apply computer models (GIS in Environmental Contamination) for analyzing transport and fate of soil chemicals. Various technologies have been developed for remediation of oil-contaminated soil/ sediments There are several principal strategies for remediation:

- Excavate soil and take it to a disposal site away from ready pathways for human or sensitive ecosystem contact. This technique also applies to dredging of bay muds containing toxins.
- Aeration of soils at the contaminated site (with attendant risk of creating air pollution).
- Thermal remediation by introduction of heat to raise subsurface temperatures sufficiently high to volatize chemical contaminants out of the soil for vapour extraction. Technologies include ISTD, electrical resistance heating (ERH), and ET-DSP[tm].
- Bioremediation, involving microbial digestion of certain organic chemicals. Techniques used in bioremediation include landfarming, biostimulation and bioaugmentating soil biota with commercially available microflora.
- Extraction of groundwater or soil vapor with an active electromechanical system, with subsequent stripping of the contaminants from the extract.
- Containment of the soil contaminants (such as by capping or paving over in place).
- Phytoremediation, or using plants (such as willow) to extract heavy metals.
- Mycoremediation, or using fungus to metabolize contaminants and accumulate heavy metals.

By Country

Various national standards for concentrations of particular contaminants include the United States EPA Region 9 Preliminary Remediation Goals (U.S. PRGs), the U.S.

EPA Region 3 Risk Based Concentrations (U.S. EPA RBCs) and National Environment Protection Council of Australia Guideline on Investigation Levels in Soil and Groundwater.

People's Republic of China

The immense and sustained growth of the People's Republic of China since the 1970s has exacted a price from the land in increased soil pollution. The State Environmental Protection Administration believes it to be a threat to the environment, to food safety and to sustainable agriculture. According to a scientific sampling, 150 million mi (100,000 square kilometers) of China's cultivated land have been polluted, with contaminated water being used to irrigate a further 31.5 million mi (21,670 km^2.) and another 2 million mi (1,300 square kilometers) covered or destroyed by solid waste. In total, the area accounts for one-tenth of China's cultivatable land, and is mostly in economically developed areas. An estimated 12 million tonnes of grain are contaminated by heavy metals every year, causing direct losses of 20 billion yuan (US$2.57 billion).

United Kingdom

Generic guidance commonly used in the United Kingdom are the Soil Guideline Values published by the Department for Environment, Food and Rural Affairs (DEFRA) and the Environment Agency. These are screening values that demonstrate the minimal acceptable level of a substance. Above this there can be no assurances in terms of significant risk of harm to human health. These have been derived using the Contaminated Land Exposure Assessment Model (CLEA UK). Certain input parameters such as Health Criteria Values, age and land use are fed into CLEA UK to obtain a probabilistic output.

Guidance by the Inter Departmental Committee for the Redevelopment of Contaminated Land (ICRCL) has been formally withdrawn by DEFRA, for use as a prescriptive document to determine the potential need for remediation or further assessment.

The CLEA model published by DEFRA and the Environment Agency (EA) in March 2002 sets a framework for the appropriate assessment of risks to human health from contaminated land, as required by Part IIA of the Environmental Protection Act 1990. As part of this framework, generic Soil Guideline Values (SGVs) have currently been derived for ten contaminants to be used as "intervention values". These values should not be considered as remedial targets but values above which further detailed assessment should be considered.

Three sets of CLEA SGVs have been produced for three different land uses, namely

- residential (with and without plant uptake)
- allotments
- commercial/industrial

It is intended that the SGVs replace the former ICRCL values. It should be noted that the CLEA SGVs relate to assessing chronic (long term) risks to human health and do not apply to the protection of ground workers during construction, or other potential receptors such as groundwater, buildings, plants or other ecosystems. The CLEA SGVs are not directly applicable to a site completely covered in hardstanding, as there is no direct exposure route to contaminated soils.

To date, the first ten of fifty-five contaminant SGVs have been published, for the following: arsenic, cadmium, chromium, lead, inorganic mercury, nickel, selenium ethyl benzene, phenol and toluene. Draft SGVs for benzene, naphthalene and xylene have been produced but their publication is on hold. Toxicological data (Tox) has been published for each of these contaminants as well as for benzo[a]pyrene, benzene, dioxins, furans and dioxin-like PCBs, naphthalene, vinyl chloride, 1,1,2,2 tetrachloroethane and 1,1,1,2 tetrachloroethane, 1,1,1 trichloroethane, tetrachloroethene, carbon tetrachloride, 1,2-dichloroethane, trichloroethene and xylene. The SGVs for ethyl benzene, phenol and toluene are dependent on the soil organic matter (SOM) content (which can be calculated from the total organic carbon (TOC) content). As an initial screen the SGVs for 1% SOM are considered to be appropriate.

Canada

India

In March 2009, the issue of Uranium poisoning in Punjab attracted press coverage. It was alleged to be caused by fly ash ponds of thermal power stations, which reportedly lead to severe birth defects in children in the Faridkot and Bhatinda districts of Punjab. The news reports claimed the uranium levels were more than 60 times the maximum safe limit. In 2012, the Government of India confirmed that the ground water in Malwa belt of Punjab has uranium metal that is 50% above the trace limits set by the United Nations' World Health Organization. Scientific studies, based on over 1000 samples from various sampling points, could not trace the source to fly ash and any sources from thermal power plants or industry as originally alleged. The study also revealed that the uranium concentration in ground water of Malwa district is not 60 times the WHO limits, but only 50% above the WHO limit in 3 locations. This highest concentration found in samples was less than those found naturally in ground waters currently used for human purposes elsewhere, such as Finland. Research is underway to identify natural or other sources for the uranium.

Land Degradation

Serious land degradation in Nauru after the depletion of the phosphate cover through mining.

Land degradation is a process in which the value of the biophysical environment is affected by a combination of human-induced processes acting upon the land. It is viewed as any change or disturbance to the land perceived to be deleterious or undesirable. Natural hazards are excluded as a cause; however human activities can indirectly affect phenomena such as floods and bush fires.

This is considered to be an important topic of the 21st century due to the implications land degradation has upon agronomic productivity, the environment, and its effects on food security. It is estimated that up to 40% of the world's agricultural land is seriously degraded.

Measures

Land degradation is a broad term that can be applied differently across a wide range of scenarios. There are four main ways of looking at land degradation and its impact on the environment around it:

1. A temporary or permanent decline in the productive capacity of the land. This can be seen through a loss of biomass, a loss of actual productivity or in potential productivity, or a loss or change in vegetative cover and soil nutrients.

2. Action in the land's capacity to provide resources for human livelihoods. This can be measured from a base line of past land use.

3. Loss of biodiversity: A loss of range of species or ecosystem complexity as a decline in the environmental quality.

4. Shifting ecological risk: increased vulnerability of the environment or people to destruction or crisis. This is measured through a base line in the form of pre-existing risk of crisis or destruction.

A problem with defining land degradation is that what one group of people might view as degradation, others might view as a benefit or opportunity. For example, planting crops at a location with heavy rainfall and steep slopes would create scientific and environmental concern regarding the risk of soil erosion by water, yet farmers could view the location as a favourable one for high crop yields.

Different Types

In addition to the usual types of land degradation that have been known for centuries (water, wind and mechanical erosion, physical, chemical and biological degradation), four other types have emerged in the last 50 years:

- pollution, often chemical, due to agricultural, industrial, mining or commercial activities;
- loss of arable land due to urban construction;
- artificial radioactivity, sometimes accidental;
- land-use constraints associated with armed conflicts.

Overall, 36 types of land degradation can be assessed. All are induced or aggravated by human activities, e.g. sheet erosion, silting, aridification, salinization, urbanization, etc.

Causes

Overgrazing by livestock can lead to land degradation.

Land degradation is a global problem largely related to agricultural use. The major causes include:

- Land clearance, such as clearcutting and deforestation
- Agricultural depletion of soil nutrients through poor farming practices
- Livestock including overgrazing and overdrafting
- Inappropriate irrigation and overdrafting
- Urban sprawl and commercial development
- Vehicle off-roading
- Quarrying of stone, sand, ore and minerals
- Increase in field size due to economies of scale, reducing shelter for wildlife, as hedgerows and copses disappear
- Exposure of naked soil after harvesting by heavy equipment
- Monoculture, destabilizing the local ecosystem
- Dumping of non-biodegradable trash, such as plastics
- Soil degradation, e.g.
    - Soil contamination
    - Soil erosion
    - Soil acidification
    - Loss of soil carbon

Effects

Overcutting of vegetation occurs when people cut forests, woodlands and shrublands—to obtain timber, fuelwood and other products—at a pace exceeding the rate of natural regrowth. This is frequent in semi-arid environments, where fuelwood shortages are often severe.

Overgrazing is the grazing of natural pastures at stocking intensities above the livestock carrying capacity; the resulting decrease in the vegetation cover is a leading cause of wind and water erosion. It is a significant factor in Afghanistan. The growing population pressure, during 1980-1990, has led to decreases in the already small areas of agricultural land per person in six out of eight countries (14% for India and 22% for Pakistan).

Soil erosion in a wheat field near Pullman, USA.

Population pressure also operates through other mechanisms. Improper agricultural practices, for instance, occur only under constraints such as the saturation of good lands under population pressure which leads settlers to cultivate too shallow or too steep soils, plough fallow land before it has recovered its fertility, or attempt to obtain multiple crops by irrigating unsuitable soils.

High population density is not always related to land degradation. Rather, it is the practices of the human population that can cause a landscape to become degraded. Populations can be a benefit to the land and make it more productive than it is in its natural state. Land degradation is an important factor of internal displacement in many African and Asian countries.

Severe land degradation affects a significant portion of the Earth's arable lands, decreasing the wealth and economic development of nations. As the land resource base becomes less productive, food security is compromised and competition for dwindling resources increases, the seeds of famine and potential conflict are sown.

Sensitivity and Resilience

Sensitivity and resilience are measures of the vulnerability of a landscape to degradation. These two factors combine to explain the degree of vulnerability. Sensitivity is the degree to which a land system undergoes change due to natural forces, human intervention or a combination of both. Resilience is the ability of a landscape to absorb change, without significantly altering the relationship between the relative importance and numbers of individuals and species that compose the community. It also refers to the ability of the region to return to its original state after being changed in some way. The resilience of a landscape can be increased or decreased through human interaction based upon different methods of land-use management. Land that is degraded becomes less resilient than undegraded land, which can lead to even further degradation through shocks to the landscape.

Climate Change

Significant land degradation from seawater inundation, particularly in river deltas and on low-lying islands, is a potential hazard that was identified in a 2007 IPCC report.

IPCC
Assessment reports:
First (1990)
1992 sup.
Second (1995)
Third (2001)
Fourth (2007)
Fifth (2014)
UNFCCC | WMO | UNEP

As a result of sea-level rise from climate change, salinity levels can reach levels where agriculture becomes impossible in very low-lying areas.

Soil Retrogression and Degradation

Soil retrogression and degradation are two regressive evolution processes associated with the loss of equilibrium of a stable soil. Retrogression is primarily due to soil erosion and corresponds to a phenomenon where succession reverts the land to its natural physical state. Degradation is an evolution, different from natural evolution, related to the local climate and vegetation. It is due to the replacement of primary plant communities (known as climax vegetation) by the secondary communities. This replacement modifies the humus composition and amount, and affects the formation of the soil. It is directly related to human activity. Soil degradation may also be viewed as any change or ecological disturbance to the soil perceived to be deleterious or undesirable.

General

At the beginning of soil formation, the bare rock out crops is gradually colonized by pioneer species (lichens and mosses). They are succeeded by herbaceous vegetation, shrubs and finally forest. In parallel, the first humus-bearing horizon is formed (the A horizon), followed by some mineral horizons (B horizons). Each successive stage is characterized by a certain association of soil/vegetation and environment, which defines an ecosystem.

Intensive tillage result on soil degradation.

Willow hedge strengthened with fascines for the limitation of runoff, north of France.

After a certain time of parallel evolution between the ground and the vegetation, a state of steady balance is reached. This stage of development is called climax by some ecologists and "natural potential" by others. Succession is the evolution towards climax. Regardless of its name, the equilibrium stage of primary succession is the highest natural form of development that the environmental factors are capable of producing.

The cycles of evolution of soils have very variable durations, between tens, hundreds, or thousands of years for quickly evolving soils (A horizon only) to more than a million years for slowly developing soils. The same soil may achieve several successive steady state conditions during its existence, as exhibited by the Pygmy forest sequence in Mendocino County, California. Soils naturally reach a state of high productivity, from which they naturally degrade as mineral nutrients are removed from the soil system. Thus older soils are more vulnerable to the effects of induced retrogression and degradation.

Ecological Factors Influencing Soil Formation

There are two types of ecological factors influencing the evolution of a soil (through alteration and humification). These two factors are extremely significant to explain the evolution of soils of short development.

- A first type of factor is the average climate of an area and the vegetation which is associated (biome).
- A second type of factor is more local, and is related to the original rock and local drainage. This type of factor explains appearance of specialized associations (ex peat bogs).

Biorhexistasy Theory

The destruction of the vegetation implies the destruction of evoluted soils, or a regressive evolution. Cycles of succession-regression of soils follow one another within short intervals of time (human actions) or long intervals of time (climate variations).

The climate role in the deterioration of the rocks and the formation of soils lead to the formulation of the theory of the biorhexistasy.

- In wet climate, the conditions are favorable to the deterioration of the rocks (mostly chemically), the development of the vegetation and the formation of soils; this period favorable to life is called biostasy.
- In dry climate, the rocks exposed are mostly subjected to mechanical disintegration which produces coarse detrital materials: this is referred to as rhexistasy.

Perturbations of the Balance of a Soil

When the state of balance, characterized by the ecosystem climax is reached, it tends to be maintained stable in the course of time. The vegetation installed on the ground provides the humus and ensures the ascending circulation of the matters. It protects the ground from erosion by playing the role of barrier (for example, protection from water and wind). Plants can also reduce erosion by binding the particles of the ground to their roots.

A disturbance of climax will cause retrogression, but often, secondary succession will start to guide the evolution of the system after that disturbance. Secondary succession is much faster than primary because the soil is already formed, although deteriorated and needing restoration as well.

However, when a significant destruction of the vegetation takes place (of natural origin such as an avalanche or human origin), the disturbance undergone by the ecosystem is too important. In this latter case, erosion is responsible for the destruction of the upper horizons of the ground, and is at the origin of a phenomenon of reversion to pioneer conditions. The phenomenon is called retrogression and can be partial or total (in this case, nothing remains beside bare rock). For example, the clearing of an inclined ground, subjected to violent rains, can lead to the complete destruction of the soil. Man can deeply modify the evolution of the soils by direct and brutal action, such as clearing, abusive cuts, forest pasture, litters raking. The climax vegetation is gradually replaced and the soil modified (example: replacement of leafy tree forests by moors or pines plantations). Retrogression is often related to very old human practices.

Influence of Human Activity

Soil erosion is the main factor for soil degradation and is due to several mechanisms: water erosion, wind erosion, chemical degradation and physical degradation.

Erosion is strongly related to human activity. For example, roads which increase impermeable surfaces lead to streaming and ground loss. Agriculture also accelerates soil erosion (increase of field size, correlated to hedges and ditches removal). Meadows are in regression to the profit of plowed lands. Spring cultures (sunflower, corn, beet) surfaces are increasing and leave the ground naked in winter. Sloping grounds are gradually colonized by vine. Lastly, use of herbicides leaves the ground naked between each crop. New cultural practices, such as mechanization also increases the risks of erosion. Fertilization by mineral manures rather than organic manure gradually destructure the soil. Many scientistsobserved a gradual decrease of soil organic matter content in soils, as well as a decrease of soil biological activity (in particular, in relation to chemical uses). Lastly, deforestation, in particular, is responsible for degradation of forest soils.

Agriculture increases the risk of erosion through its disturbance of vegetation by way of:

- Overgrazing of animals
- Monoculture planting
- Row cropping
- Tilling or plowing
- Crop removal
- Land-use conversion

Consequences of Soil Regression and Degradation

- yields impact: Recent increases in the human population have placed a great strain on the world's soil systems. More than 6 billion people are now using about 38% of the land area of the Earth to raise crops and livestock. Many soils suffer from various types of degradation, that can ultimately reduce their ability to produce food resources. Slight degradation refers to land where yield potential has been reduced by 10%, moderate degradation refers to a yield decrease from 10-50%. Severely degraded soils have lost more than 50% of their potential. Most severely degraded soils are located in developing countries.

- natural disasters: natural disasters such as mud flows, floods are responsible for the death of many living beings each year.
- deterioration of the water quality: the increase in the turbidity of water and the contribution of nitrogen and of phosphorus can result in eutrophication. Soils particles in surface waters are also accompanied by agricultural inputs and by some pollutants of industrial, urban and road origin (such as heavy metals). The ecological impact of agricultural inputs (such as weed killer) is known but difficult to evaluate because of the multiplicity of the products and their broad spectrum of action.
- biological diversity: soil degradation may involve perturbation of microbial communities, disappearance of the climax vegetation and decrease in animal habitat, thus leading to a biodiversity loss and animal extinction.
- economic loss: the estimated costs for land degradation are US $40 billion per year. This figure does not take into account negative externalities such as the increased use of fertilizers, loss of biodiversity and loss of unique landscapes.

Soil Enhancement, Rebuilding and Regeneration

Problems of soil erosion can be fought, and certain practices can lead to soil enhancement and rebuilding. Even though simple, methods for reducing erosion are often not chosen because these practices outweigh the short-term benefits. Rebuilding is especially possible through the improvement of soil structure, addition of organic matter and limitation of runoff. However, these techniques will never totally succeed to restore a soil (and the fauna and flora associated to it) that took more than 1000 years to build up. Soil regeneration is the reformation of degraded soil through biological, chemical, and or physical processes.

When productivity declined in the low-clay soils of northern Thailand, farmers initially responded by adding organic matter from termite mounds, but this was unsustainable in the long-term. Scientists experimented with adding bentonite, one of the smectite family of clays, to the soil. In field trials, conducted by scientists from the International Water Management Institute in cooperation with Khon Kaen University and local farmers, this had the effect of helping retain water and nutrients. Supplementing the farmer's usual practice with a single application of 200 kg bentonite per rai (6.26 rai =

1 hectare) resulted in an average yield increase of 73%. More work showed that applying bentonite to degraded sandy soils reduced the risk of crop failure during drought years.

In 2008, three years after the initial trials, IWMI scientists conducted a survey among 250 farmers in northeast Thailand, half who had applying bentonite to their fields and half who had not. The average output for those using the clay addition was 18% higher than for non-clay users. Using the clay had enabled some farmers to switch to growing vegetables, which need more fertile soil. This helped to increase their income. The researchers estimated that 200 farmers in northeast Thailand and 400 in Cambodia had adopted the use of clays, and that a further 20,000 farmers were introduced to the new technique.

References

- Toy, Terrence J.; Foster, George R.; Renard, Kenneth G. (2002). Soil erosion : processes, prediction, measurement, and control. New York, NY: Wiley. ISBN 9780471383697
- Blanco-Canqui, Humberto; Rattan, Lal (2008). "Water erosion". Principles of soil conservation and management. Dordrecht: Springer. pp. 21–53. ISBN 9781402087097
- Borah, Deva K. et al. (2008). "Watershed sediment yield". In Garcia, Marcelo H. Sedimentation Engineering: Processes, Measurements, Modeling, and Practice. ASCE Publishing. p. 828. ISBN 978-0-7844-0814-8
- Blanco-Canqui, Humberto; Rattan, Lal (2008). "Soil and water conservation". Principles of soil conservation and management. Dordrecht: Springer. pp. 1–20. ISBN 9781402087097
- Zheng, Xiaojing & Huang, Ning (2009). Mechanics of Wind-Blown Sand Movements. Springer. pp. 7–8. ISBN 978-3-540-88253-4
- Cornelis, Wim S. (2006). "Hydroclimatology of wind erosion in arid and semi-arid environments". In D'Odorico, Paolo & Porporato, Amilcare. Dryland Ecohydrology. Springer. p. 141. ISBN 978-1-4020-4261-4
- Blanco-Canqui, Humberto; Rattan, Lal (2008). "Wind erosion". Principles of soil conservation and management. Dordrecht: Springer. pp. 54–80. ISBN 9781402087097
- Nancy D. Gordon (2004-06-01). "Erosion and Scour". Stream hydrology: an introduction for ecologists. ISBN 978-0-470-84357-4
- Wiggs, Giles F.S. (2011). "Geomorphological hazards in drylands". In Thomas, David S.G. Arid Zone Geomorphology: Process, Form and Change in Drylands. John Wiley & Sons. p. 588. ISBN 978-0-470-71076-0
- Van Beek, Rens (2008). "Hillside processes: mass wasting, slope stability, and erosion". In Norris, Joanne E. et al. Slope Stability and Erosion Control: Ecotechnological Solutions. Springer. ISBN 978-1-4020-6675-7

9

Diverse Aspects of Soil Science

There are various tools and techniques used in soil science. One of them is soil moisture sensor. It is a device which is used to measure the water content present in the soil. This chapter encompasses the tools and techniques related to soil science such as soil nailing, soil compaction, soil bioengineering and pedometrics. The topics elaborated in this chapter will help in gaining a better perspective about these techniques.

Soil Moisture Sensor

A simple soil moisture sensor for gardeners.

Soil moisture sensors measure the volumetric water content in soil. Since the direct gravimetric measurement of free soil moisture requires removing, drying, and weighting of a sample, soil moisture sensors measure the volumetric water content indirectly by using some other property of the soil, such as electrical resistance, dielectric constant, or interaction with neutrons, as a proxy for the moisture content. The relation between the measured property and soil moisture must be calibrated and may vary depending on environmental factors such as soil type, temperature, or electric conductivity. Reflected microwave radiation is affected by the soil moisture and is used for remote sensing in hydrology and agriculture. Portable probe instruments can be used by farmers or gardeners.

Soil moisture sensors typically refer to sensors that estimate volumetric water content. Another class of sensors measure another property of moisture in soils called water potential; these sensors are usually referred to as soil water potential sensors and include tensiometers and gypsum blocks.

Technology

Technologies commonly used to indirectly measure volumetric water content (soil moisture) include:

- Frequency Domain Reflectometry (FDR): The dielectric constant of a certain volume element around the sensor is obtained by measuring the operating frequency of an oscillating circuit.
- Time Domain Transmission (TDT) and Time Domain Reflectometry (TDR): The dielectric constant of a certain volume element around the sensor is obtained by measuring the speed of propagation along a buried transmission line.
- Neutron moisture gauges: The moderator properties of water for neutrons are utilized to estimate soil moisture content between a source and detector probe.
- Soil resistivity: Measuring how strongly the soil resists the flow of electricity between two electrodes can be used to determine the soil moisture content.
- Galvanic cell: The amount of water present can be determined based on the voltage the soil produces because water acts as an electrolyte and produces electricity. The technology behind this concept is the galvanic cell.

Application

Agriculture

Measuring soil moisture is important for agricultural applications to help farmers manage their irrigation systems more efficiently. Knowing the exact soil moisture conditions on their fields, not only are farmers able to generally use less water to grow a crop, they are also able to increase yields and the quality of the crop by improved management of soil moisture during critical plant growth stages.

Landscape Irrigation

In urban and suburban areas, landscapes and residential lawns are using soil moisture sensors to interface with an irrigation controller. Connecting a soil moisture sensor to a simple irrigation clock will convert it into a “smart” irrigation controller that prevents irrigation cycles when the soil is already wet, e.g. following a recent rainfall event.

Golf courses are using soil moisture sensors to increase the efficiency of their irrigation systems to prevent over-watering and leaching of fertilizers and other chemicals into the ground.

Research

Soil moisture sensors are used in numerous research applications, e.g. in agricultural science and horticulture including irrigation planning, climate research, or environmental science including solute transport studies and as auxiliary sensors for soil respiration measurements.

Simple Sensors for Gardeners

Relatively cheap and simple devices that do not require a power source are available for checking whether plants have sufficient moisture to thrive. After inserting a probe into the soil for approximately 60 seconds a meter indicates if the soil is too dry, moist or wet for plants.

Soil Nailing

Soil nailing is a construction technique that can be used as a remedial measure to treat unstable natural soil slopes or as a construction technique that allows the safe over-steepening of new or existing soil slopes. The technique involves the insertion of relatively slender reinforcing elements into the slope – often general purpose reinforcing bars (rebar) although proprietary solid or hollow-system bars are also available. Solid bars are usually installed into pre-drilled holes and then grouted into place using a separate grout line, whereas hollow bars may be drilled and grouted simultaneously by the use of a sacrificial drill bit and by pumping grout down the hollow bar as drilling progresses. Kinetic methods of firing relatively short bars into soil slopes have also been developed. Bars installed using drilling techniques are usually fully grouted and installed at a slight downward inclination with bars installed at regularly spaced points across the slope face. A rigid facing (often pneumatically applied concrete, otherwise known as shotcrete) or isolated soil nail head plates may be used at the surface. Alternatively a flexible reinforcing mesh may be held against the soil face beneath the head plates. Rabbit proof wire mesh and environmental erosion control fabrics and may be used in conjunction with flexible mesh facing where environmental conditions dictate.

Soil nail components may also be used to stabilize retaining walls or existing fill slopes (embankments and levees); this is normally undertaken as a remedial measure.

Since its first application using modern techniques in Versailles in 1972, soil nailing is now a well-established technique around the world. One of the first national guideline publications for soil nailing was produced in Japan in 1987. The U.S. Federal Highway Administration issued guideline publications in 1996 and 2003.

Preliminary Analysis

Four main points to be considered in determining if soil nailing would be an effective retention technique are as follows. First, the existing ground conditions should be examined. Next, the advantages and disadvantages for a soil nail wall should be assessed for the particular application being considered. Then other systems should be considered for the particular application. Finally, cost of the soil nail wall should be considered. Soil nail walls can be used for a variety of soil types and conditions. The most favorable conditions for soil nailing are as follows: The soil should be able to stand unsupported one to two meters high for a minimum of two days when cut vertical or nearly vertical. Also all soil nails within a cross section should be located above the groundwater table. If the soil nails are not located above the groundwater table, the groundwater should not negatively affect the face of the excavation, the bond between the ground and the soil nail itself. Based upon these favorable conditions for soil nailing stiff to hard fine-grained soils which include stiff to hard clays, clayey silts, silty clays, sandy clays, and sandy silts are preferred soils. Sand and gravels which are dense to very dense soils with some apparent cohesion also work well for soil nailing. Weathered rock is also acceptable as long as the rock is weathered evenly throughout (meaning no weakness planes). Finally, glacial soils work well for soil nailing.

A list of unfavorable or difficult soil conditions for soil nailing can include dry, poorly graded cohesion-less soils, soils with a high groundwater table, soils with cobbles and boulders, soft to very soft fine-grained soils, highly corrosive soils, weathered rock with unfavorable weakness planes, and loess. Other difficult conditions include prolonged exposure to freezing temperatures, a climate that has a repeated freeze-and-thaw cycle, and granular soils that are very loose.

Origins

Soil nailing evolved from the New Austrian Tunneling method which is a system for underground excavations in rock. This method consists of passive steel reinforcement in the rock followed by the application of reinforced shotcrete. This concept of combining passive steel reinforcement and shotcrete has also been applied to the stabilization of rock slopes since the early 1960s.

The first application of soil nailing was implemented in 1972 for a railroad widening project near Versailles, France. Soil nails were used to stabilize an 18 m high slope consisting of sandy soil. This method proved to be more cost-effective, while at the same time cut down the construction time when compared to other conventional support methods. Germany was the next country to investigate soil nailing. From 1975 to 1981 the University of Karlsruhe and the construction company Bauer collaborated to establish a research program. This program conducted full scale testing of experimental walls with different configurations and developed analysis procedures for use in design. The United States first used soil nailing in 1976 for the support of a 13.7 m deep foundation excavation in dense silty sands. Soil nailing was implemented in the expansion of The Good Samaritan Hospital in Portland, Oregon. This retaining system was produced in approximately half the time at about 85% of the cost of conventional retaining systems.

Design

After a preliminary analysis of the site, initial designs of the soil nail wall can be begin. This begins with a selection of limit states and design approaches. The two most common limit states used in soil nail wall design is strength limit and service limit states. The strength limit state is the limit state that addresses potential failure mechanisms or collapse states of the soil nail wall system. The service limit state is the limit state that addresses loss of service function resulting from excessive wall deformation and is defined by restrictions in stress, deformation and facing crack width under regular service conditions. The two most common design approaches for soil nail walls are limit state design and service load design.

Initial design considerations include wall layout (wall height and length), soil nail vertical and horizontal spacing, soil nail pattern on wall face, soil nail inclination, soil nail length and distribution, soil nail material and relevant ground properties. With all these variables in the mind of the design engineer the next step is to use simplified charts to preliminarily evaluate nail length and maximum nail force. Nail length, diameter and spacing typically control external and internal stability of the wall. These parameters can be adjusted during design until all external and internal stability requirements are met. After the initial design is completed, final design progresses where the soil nail wall has to be tested for external and internal failure modes, seismic considerations and aesthetic qualities. Drainage, frost penetration and external loads such as wind and hydrostatic forces also have to be determined and included in the final examination of the design. Soil nail walls are not ideal in locations with highly plastic clay soils. Soils with high plasticity, a high liquid limit and low undrained shear strengths are at risk of long-term deformation (creep).

Construction

With the design complete, construction is the next step. Most soil nail wall construction follows a specific procedure. First a cut is excavated and temporary bracing is put in place if necessary. This is done with conventional earth moving equipment and hydraulic drills. Next, holes for the soil nails are drilled at predetermined locations as specified by the design engineer. The equipment used for this step is dependent on the stability of the material in which the soil nail wall is supporting. Rotary or rotary percussive methods using air flush or dry auger methods can be used with stable ground. For unstable ground, single tube and duplex rotary methods with air and water flush or hollow stem auger methods are used. With the holes drilled, the next step is to install and grout the nails into place. After all nails are inserted, a drainage system is put into place. Synthetic drainage mat is placed vertically between the nail heads, which are extended down to the base of the wall where they are most commonly connected to a footing drain. A layer of shotcrete is applied and bearing plates are installed before a final facing is put in place to complete the soil nail wall. Variations of the steps described above may be necessary to accommodate additional preparation tasks or supplementary activities for specific project conditions.

In terms of construction, soil nail walls have a decisive advantage over other alternatives. Soil nail walls require a smaller right-of-way than ground anchor walls and have less environmental impact. Installation of soil nail walls is relatively rapid and typically uses less materials and smaller construction equipment than ground anchor walls.

Cost Comparison

One great advantage of soil nail walls is their cost-effectiveness over other alternatives. When conventional soil nailing construction procedures are used, soil nail walls are much more economical than concrete gravity walls and similarly or more cost effective than ground anchor walls.

Inspection and Performance Monitoring

Inspection activities play a vital role in the production of high-quality soil nail walls because conformance to project plans and specifications should result in a soil nail wall that will perform its intended duty for its designed duration. Inspections usually involve evaluation of the following: conformance of system components to material specification, conformance of construction methods to execution specifications, conformance to short-term performance specifications, and long-term monitoring. Short-term performance specifications are checked with loads tests, which utilize hydraulic jacks and pumps to perform several load applications. Three common load tests for short-term performance are verification or ultimate load tests, proof tests and creep tests. Verification or ultimate load tests are conducted to verify the compliance of the soil nails with pullout capacity and strengths resulting from the contractor's installation method. Proof tests are intended to verify that the contractor's construction procedure has been consistent and that the nails have not been drilled and grouted in a soil zone not tested in the verification stage. Creep tests are performed to ensure that the nail design loads can be safely carried throughout the structure's service life.

Long-term performance monitoring is used to collect data to ensure adequate performance and refine future design practices. Parameters to be measured include vertical and horizontal movement of the wall face, local movements or deterioration of facing elements, drainage to the ground, loads, load distribution and load changes in the nails, temperature and rainfall. These parameters are measured using several specific tools including inclinometers, load cells and strain gauges.

Soil Compaction

In geotechnical engineering, soil compaction is the process in which a stress applied to a soil causes densification as air is displaced from the pores between the soil grains. When stress is applied that causes densification due to water (or other liquid) being displaced from between the soil grains, then consolidation, not compaction, has occurred. Normally, compaction is the result of heavy machinery compressing the soil, but it can also occur due to the passage of (e.g.) animal feet.

In soil science and agronomy, soil compaction is usually a combination of both engineering compaction and consolidation, so may occur due to a lack of water in the soil, the applied stress being internal suction due to water evaporation as well as due to passage of animal feet. Affected soils become less able to absorb rainfall, thus increasing runoff and erosion. Plants have difficulty in compacted soil because the mineral grains are pressed together, leaving little space for air and water, which are essential for root growth. Burrowing animals also find it a hostile environment, because the denser soil is more difficult to penetrate. The ability of a soil to recover from this type of compaction depends on climate, mineralogy and fauna. Soils with high shrink-swell capacity, such as vertisols, recover quickly from compaction where moisture conditions are variable (dry spells shrink the soil, causing it to crack). But clays which do not crack as they dry cannot recover from compaction on their own unless they host ground-dwelling animals such as earthworms — the Cecil soil series is an example.

A crawler-backhoe is here equipped with a narrow sheepsfoot roller to compact the fill over newly placed sewer pipe, forming a stable support for a new road surface.

A compactor/roller fitted with a sheepsfoot drum.

A Hamm vibrating roller with plain drum as used for compacting asphalt and granular soils.

Wacker Neuson vibratory rammer BS 60-2i in action.

In Construction

Soil compaction is a vital part of the construction process. It is used for support of structural entities such as building foundations, roadways, walkways, and earth retaining structures to name a few. For a given soil type certain properties may deem it more or less desirable to perform adequately for a particular circumstance. In general, the preselected soil should have adequate strength, be relatively incompressible so that future settlement is not significant, be stable against volume change as water content or other factors vary, be durable and safe against deterioration, and possess proper permeability.

When an area is to be filled or backfilled the soil is placed in layers called lifts. The ability of the first fill layers to be properly compacted will depend on the condition of the natural material being covered. If unsuitable material is left in place and backfilled, it may compress over a long period under the weight of the earth fill, causing settlement cracks in the fill or in any structure supported by the fill. In order to determine if the natural soil will support the first fill layers, an area can be proofrolled. Proofrolling consists of utilizing a piece heavy construction equipment (typically, heavy compaction equipment or hauling equipment) to roll across the fill site and watching for deflections to be revealed. These areas will be indicated by the development of rutting, pumping, or ground weaving.

To ensure adequate soil compaction is achieved, project specifications will indicate the required soil density or degree of compaction that must be achieved. These specifications are generally recommended by a geotechnical engineer in a geotechnical engineering report.

The soil type - that is, grain-size distributions, shape of the soil grains, specific gravity of soil solids, and amount and type of clay minerals, present - has a great influence on the maximum dry unit weight and optimum moisture content. It also has a great influence on how the materials should be compacted in given situations. Compaction is accomplished by use of heavy equipment. In sands and gravels, the equipment usually vibrates, to cause re-orientation of the soil particles into a denser configuration. In silts and clays, a sheepsfoot roller is frequently used, to create small zones of intense shearing, which drives air out of the soil.

Determination of adequate compaction is done by determining the in-situ density of the soil and comparing it to the maximum density determined by a laboratory test. The most commonly used laboratory test is called the Proctor compaction test and there are two different methods in obtaining the maximum density. They are the standard Proctor and modified Proctor tests; the modified Proctor is more commonly used. For small dams, the standard Proctor may still be the reference.

While soil under structures and pavements needs to be compacted, it is important after construction to decompact areas to be landscaped so that vegetation can grow.

Compaction Methods

There are several means of achieving compaction of a material. Some are more appropriate for soil compaction than others, while some techniques are only suitable for particular soils or soils in particular conditions. Some are more suited to compaction of non-soil materials such as asphalt. Generally, those that can apply significant amounts of shear as well as compressive stress, are most effective.

The available techniques can be classified as:

1. Static - a large stress is slowly applied to the soil and then released.
2. Impact - the stress is applied by dropping a large mass onto the surface of the soil.
3. Vibrating - a stress is applied repeatedly and rapidly via a mechanically driven plate or hammer. Often combined with rolling compaction.
4. Gyrating - a static stress is applied and maintained in one direction while the soil is a subjected to a gyratory motion about the axis of static loading. Limited to laboratory applications.
5. Rolling - a heavy cylinder is rolled over the surface of the soil. Commonly used on sports pitches. Roller-compactors are often fitted with vibratory devices to enhance their effectiveness.
6. Kneading - shear is applied by alternating movement in adjacent positions. An example, combined with rolling compaction, is the 'sheepsfoot' roller used in waste compaction at landfills.

The construction plant available to achieve compaction is extremely varied and is described elsewhere.

Test Methods in Laboratory

Soil compactors are used to perform test methods which cover laboratory compaction methods used to determine the relationship between molding water content and dry unit weight of soils. Soil placed as engineering fill is compacted to a dense state to obtain satisfactory engineering properties such as, shear strength, compressibility, or permeability. In addition, foundation soils are often compacted to improve their engineering properties. Laboratory compaction tests provide the basis for determining the percent compaction and molding water content needed to achieve the required engineering properties, and for controlling construction to assure that the required compaction and water contents are achieved. Test methods such as EN 13286-2, EN 13286-47, ASTM D698, ASTM D1557, AASHTO T99, AASHTO T180, AASHTO T193, BS 1377:4 provide soil compaction testing procedures.

Soil Bioengineering

Soil bioengineering is a discipline of civil engineering. It pursues technological, ecological, economic as well as design goals and seeks to achieve these primarily by making use of living materials, i.e. seeds, plants, part of plants and plant communities, and employing them in near–natural constructions while exploiting the manifold abilities inherent in plants. Soil bioengineering may sometimes be a substitute for classical engineering works; however, in most cases it is a meaningful and necessary method of complementing the latter. Its application suggests itself in all fields of soil and hydraulic engineering, especially for slope and embankment stabilization and erosion control.

Soil bioengineering is the use of living plant materials to provide some engineering function. Soil bioengineering is an effective tool for treatment of a variety of unstable and / or eroding sites. Soil bioengineering techniques have been used for many centuries. More recently Schiechtl (1980) has encouraged the use of soil bioengineering with a variety of European examples. Soil bioengineering is now widely practiced throughout the world for the treatment of erosion and unstable slopes.

Fields of Application and Plants for Soil Bioengineering Control Works

Soil Bioengineering methods can be applied wherever the plants which are used as living building materials are able to grow well and develop. This is the case in tropical, subtropical and temperate zones whereas there are obvious limits in dry and cold regions, i.e. where arid, semi–arid and frost zones prevail. In exceptional cases, lack of water may be compensated for by watering or irrigation. In Europe, dry conditions limiting application exist in the Mediterranean as well as in some inner alpine and eastern European snowy regions. However, limits are most frequently imposed in alpine and arctic regions. These can usually be clearly noticed by the limited growth of woody plants (forest, tree and shrub lines) and the upper limits of closed turf cover. The more impoverished a region is in species, the less suited it is for the application of bioengineering methods.

Functions and Effects of Soil Bioengineering Structures

Technical Functions

- Protection of soil surface from erosion by wind, precipitation, frost and flowing water.
- Protection from rock fall.
- Elimination or binding of destructive mechanical forces.
- Reduction of flow velocity along banks.
- Surface and/or deep soil cohesion and stabilization.
- Drainage.

- Protection from wind.
- Aiding the deposition of snow, drift sand and sediments.
- Increasing soil roughness and thus preventing avalanche release.

Apart from these, ecological functions are gaining in importance, particularly as these can be fulfilled to a very limited extent only by classical engineering constructions.

Ecological Functions

- Improvement of water regime by improved soil interception and storage capability as well as water.
- Consumption by plants.
- Soil drainage.
- Protection from wind.
- Protection from ambient air pollution.
- mechanical soil amelioration by the roots of plants.
- Balancing of temperature conditions in near–ground layers of air and in the soil.
- Shading.
- Improvement of nutrient content in the soil and thus of soil fertility on previously raw soils.
- Balancing of snow deposits.
- Noise protection.
- Yield increase on neighbouring cropland.

Landscaping Functions

- Healing of wounds inflicted on the landscape by disasters and humans (exploitation of mineral resources, construction work, deposition of overburden, tunnel excavation material, industrial and domestic waste).
- Integration of structures into the landscape.
- Concealment of offending structures.
- Enrichment of the landscape by creating new features and structures, shapes and colours of vegetation.

Economic Effects

Bioengineering control works are not always necessarily cheaper in construction when compared to classical engineering structures. However, when taking into account their lifetime including their service and maintenance, they will normally turn out to be more economical. Their special advantages are:

- Lower construction costs compared to "hard" constructions.
- Lower maintenance and rehabilitation costs.
- Creation of useful green areas and woody plant populations on previously derelict land.
- Useful for income generation.

The result of soil bioengineering protection works are living systems which develop further and maintain their balance by natural succession (i.e. by dynamic self–control, without artificial input of energy). If the right living but also non–living building materials and the appropriate types of construction are chosen, exceptionally high sustainability requiring little maintenance effort can be achieved.

Pedometrics

Pedometrics is the application of mathematical and statistical methods for the study of the distribution and genesis of soils.

Pedometrics is a portmanteau of the Greek roots *pedos* (soil) and *metron* (measurement). Measurement in this case is restricted to mathematical and statistical methods as it relates to pedology, the branch of soil science that studies soil in its natural setting.

Pedometrics addresses soil related problems when there is uncertainty due to deterministic or stochastic variation, vagueness and lack of knowledge of soil properties and processes. It relies upon mathematical, statistical, and numerical methods and includes numerical approaches to classification to deal with a supposed deterministic variation. Simulation models incorporate uncertainty by adopting chaos theory, statistical distribution, or fuzzy logic.

Pedometrics addresses pedology from the perspective of emerging scientific fields such as wavelets analysis, fuzzy set theory and data mining in soil data modelling applications. The advance of pedometrics is also linked to improvements in remote and close-range sensing.

Pedometrics society logo.

References

- McCarthy, David F. (2007). Essentials of Soil Mechanics and Foundations. Upper Saddle River, NJ: Pearson Prentice Hall. p. 602. ISBN 0-13-114560-6
- McCarthy, David F. (2007). Essentials of Soil Mechanics and Foundations. Upper Saddle River, NJ: Pearson Prentice Hall. pp. 601–602. ISBN 0-13-114560-6
- McCarthy, David F. (2007). Essentials of Soil Mechanics and Foundations. Upper Saddle River, NJ: Pearson Prentice Hall. p. 595. ISBN 0-13-114560-6
- Blonquist, J. M. (2005). "A time domain transmission sensor with TDR performance characteristics" (PDF). Journal of Hydrology. 314: 235–245. doi:10.1016/j.jhydrol.2005.04.005. Retrieved 09, February 2020

Permissions

All chapters in this book are published with permission under the Creative Commons Attribution Share Alike License or equivalent. Every chapter published in this book has been scrutinized by our experts. Their significance has been extensively debated. The topics covered herein carry significant information for a comprehensive understanding. They may even be implemented as practical applications or may be referred to as a beginning point for further studies.

We would like to thank the editorial team for lending their expertise to make the book truly unique. They have played a crucial role in the development of this book. Without their invaluable contributions this book wouldn't have been possible. They have made vital efforts to compile up to date information on the varied aspects of this subject to make this book a valuable addition to the collection of many professionals and students.

This book was conceptualized with the vision of imparting up-to-date and integrated information in this field. To ensure the same, a matchless editorial board was set up. Every individual on the board went through rigorous rounds of assessment to prove their worth. After which they invested a large part of their time researching and compiling the most relevant data for our readers.

The editorial board has been involved in producing this book since its inception. They have spent rigorous hours researching and exploring the diverse topics which have resulted in the successful publishing of this book. They have passed on their knowledge of decades through this book. To expedite this challenging task, the publisher supported the team at every step. A small team of assistant editors was also appointed to further simplify the editing procedure and attain best results for the readers.

Apart from the editorial board, the designing team has also invested a significant amount of their time in understanding the subject and creating the most relevant covers. They scrutinized every image to scout for the most suitable representation of the subject and create an appropriate cover for the book.

The publishing team has been an ardent support to the editorial, designing and production team. Their endless efforts to recruit the best for this project, has resulted in the accomplishment of this book. They are a veteran in the field of academics and their pool of knowledge is as vast as their experience in printing. Their expertise and guidance has proved useful at every step. Their uncompromising quality standards have made this book an exceptional effort. Their encouragement from time to time has been an inspiration for everyone.

The publisher and the editorial board hope that this book will prove to be a valuable piece of knowledge for students, practitioners and scholars across the globe.

Index

www.ingramcontent.com/pod-product-compliance
Lightning Source LLC
LaVergne TN
LVHW080715210926
848485LV00003B/471